Practical Guide
To
Digital Electronic Circuits

FREDRICK W. HUGHES

Parker Publishing Company, Inc.
West Nyack, New York

This book is dedicated to my wife Roberta, for the many tiring hours of typing and proofreading; to my son, Jeffery for the many fishing trips missed; and to my daughter Karen, for the many good times not had during the preparation of this book.

Parker Publishing Company, Inc.
West Nyack, New York

Library of Congress Cataloging in Publication Data

Hughes, Frederick W
Practical guide to digital electronic circuits

Includes index.
1. Digital electronics. I. Title.
TK7868.D5H83 621.381 77-1855
ISBN 0-13-690735-0

Printed in the United States of America

How This Book Will Help the Technician

Digital electronic circuits are rapidly finding their way into every aspect of electronics. The electronic technician or experimenter who does not have a working knowledge of digital circuits is severely handicapped, with so many of these circuits around today, and the technician who is experienced in digital circuits will find this a lucrative area in which to apply himself. Many "want ads" today specify experience in digital circuits. Moreover, the technician, hobbyist or experimenter can build more useful and complicated devices in far less time by using digital circuits.

Most courses in digital electronics take from four months to over a year to complete and involve complicated subjects such as Boolean algebra, complex number systems and programming concepts. You don't need special pre-training to master digital electronics as it is presented in this book. Most persons interested in digital circuits want to apply the technology to their problems, not to design the actual solid-state circuitry of integrated circuits. Boolean algebra is used only to help explain basic logic gates while Karnough mapping and related design techniques are not mentioned. You will gain practical knowledge by knowing how to wire a shift register, a counter, a memory and a simple logic circuit in order to acquire the necessary degree of skill in working with digital electronics.

This book will help you "short circuit" the time normally involved in gaining experience in digital circuits. Emphasis is placed on doing rather than reading. It begins by providing you with essential information on digital circuits in the first chapter. Each chapter includes a project with complete instructions for assembling logic circuits using integrated circuits. Several of these projects, when completed, form a digital integrated circuit breadboard/tester with a self-contained power supply, clock, data switches and LED readouts. All of the circuits in the projects have been built and tested and will add greatly to updating and increasing the effectiveness of your electronic workshop. The parts used for these projects are very inexpensive and can be found in most electronic retail stores or in mail-order firms that advertise in electronic magazines. The TTL 7400-series integrated circuits used in this book are currently the mainstay of the digital electronics industry and will certainly be around for sometime. The 7400- series is pin-for-pin compatible with the newer family of CMOS integrated circuits allowing your knowledge to be transferable to a new generation of solid-state devices.

First, you will acquire knowledge and skills with the basic logic building blocks of digital circuits. You will then learn how these basic building blocks are put together to form larger building blocks for accomplishing the various and important tasks in digital circuits. There are also special sections on how integrated circuits are fabricated with the use of IC terminology and on troubleshooting these circuits. Digital circuits applications are found in the fields of computers, communications, medical equipment, high-fidelity, radio and television, electronic musical instruments, numerical control and other related areas.

The skills you will acquire from this book will save you time in building and troubleshooting digital circuits since you will know what to expect from logic circuits and digital devices. The technician who gets promotions and higher pay is the one who knows more and performs his work faster and more efficiently. You may begin cashing in on your new skills by working part-time, thus gaining more experience while making more money. The test instruments built from the projects can aid you in finding defects faster and provide you with a valuable background for using more sophisticated commercial digital test equipment.

Practical experience in working with digital circuits can be achieved in a reasonably short time by using the broad range of practical guidelines included in this book. Digital electronics is simple. In fact, digital circuits are easier to learn and construct than conventional electronic circuits. Just remember that the learning and mastery of any skill require two things; study and practice. Have scrap paper and pencil handy to be able to practice drawing the logic diagrams and data flow through a circuit. Also, constructing the projects at the end of each chapter will give you the necessary "hands-on" ability that will make you a skilled digital electronics technician.

Fredrick W. Hughes

CONTENTS

CHAPTER 1

How and Why the Binary System Is Used in Digital Circuits

Digital circuits are electronic circuits that use numerical and logical data for their operations. This data is in the form of electrical pulses, usually and ideally of the square-edge type. Originally, digital circuits were used mainly in computers where the operations dealt with numbers. More recently, these circuits are being used for monitoring and control in industrial applications as well as signal generation and modification in musical instruments and synthesizers.

Most of the world uses the decimal number system in business transactions and monetary systems. Using the decimal system for machinery and electronic equipment becomes very complex and difficult for design engineering, not to mention the difficulty of maintenance of such equipment. Therefore, a simpler system, the binary number system, is used in modern-day equipment.

In this chapter you will learn how electrical pulses are used in the binary system to operate digital equipment. You will be able to convert from one number system to the other with relative ease and also understand the similarity of arithmetic operations. Other number systems will show you how it is more efficient to communicate with digital equipment and how it is used to translate number systems which are easier to read and work with.

Project #1 involves building a simple, but very efficient, regulated power supply which is used for the other projects and testing of digital integrated circuits.

1-1 BINARY REPRESENTATION WITH VOLTAGE PULSES

The binary number system uses only the 0 and 1 digits. These *bi*nary dig*its* referred to as *bits* can be represented by electrical pulses as shown in Figure 1-1.

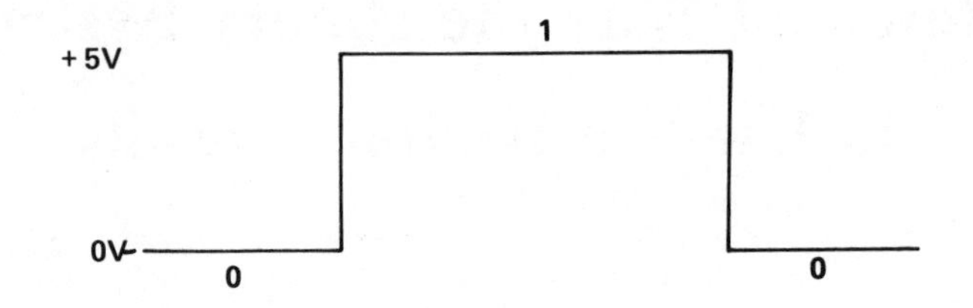

Figure 1-1. Voltage Representation of Bits

The 0 bit can be represented as zero voltage or ground (GND), while the 1 bit can be represented as a +5 volts. Digital circuits should always be in one or the other of these two states or conditions for proper operation, i.e., either high (1) or low (0). These states may also be given other names as shown below:

1	2
HIGH	LOW
POSITIVE	NEGATIVE
TRUE	FALSE
YES	NO
ON	OFF
CLOSED	OPEN

You most likely will encounter these terms in other literature on digital circuits.

A series of these pulses as shown in Figure 1-2 is referred to as a pulse train and could represent binary data in a digital circuit.

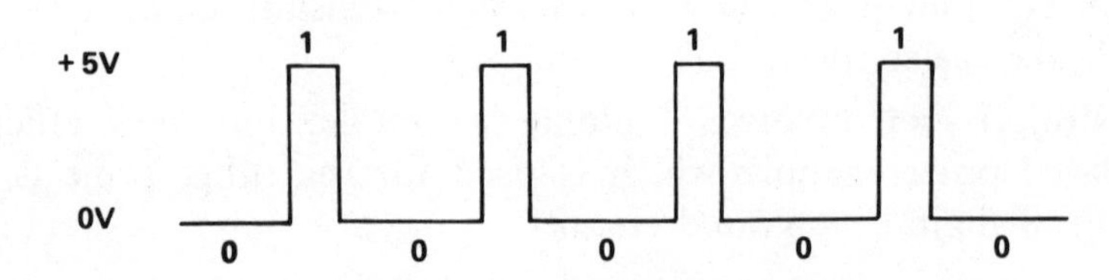

Figure 1-2. A Pulse Train

This pulse train is usually controlled by other binary signals in a circuit as illustrated in Figure 1-3.

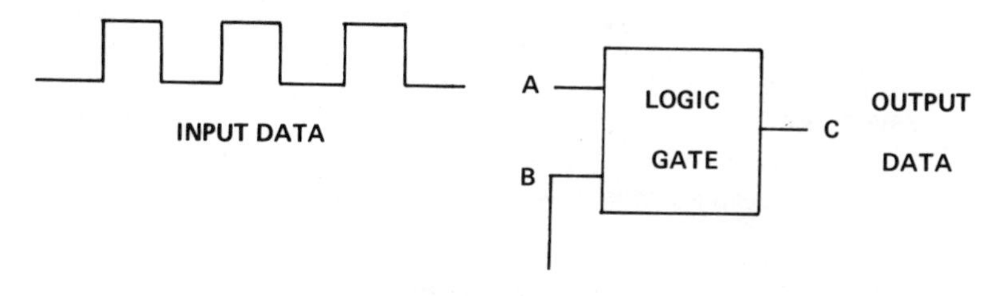

Figure 1-3. Control of Data

The block, labeled logic gate, is a generalization of the basic logic gates you will study in more detail in the next chapter. A gating signal will have to appear on input B in order to let the data on input A pass through the gate and appear at output C. For example, suppose a 1 is needed on input B to enable the gate and allow the data to pass to output C. Likewise, a 0 on input B would disable the gate and would not allow the data to pass to output C.

Since we are using pulses for data and control, it will be of value to you to know more about pulse nomenclature and definitions. Figure 1-4 illustrates the basic electrical pulse.

Figure 1-4A shows the ideal pulse. The leading edge is that part of the pulse that begins to increase in amplitude from a given reference point. The amplitude or pulse height is the maximum rise of voltage from a given reference point. The pulse width is the time duration measured between the 50- percent level in amplitude of the leading and trailing edges of a pulse. The trailing edge is that part of the pulse that begins to decrease in amplitude toward the given reference point.

Figure 1-4B shows an example of an actual pulse which is distorted because of circuit conditions. The rise time (Tr) is the time it takes a pulse to rise from 10 to 90 percent of its maximum amplitude on the leading edge. The fall time (Tf) is the time it takes a pulse to fall from 90 percent to 10 percent of its minimum amplitude on the trailing edge.

Figure 1-4C shows a time delay that might occur when a pulse is applied to a digital circuit. The time delay or turn on time is the time it takes the circuit to respond to the leading edge of the pulse. The storage time or turn off time is the time it takes a circuit to respond to the trailing edge of the pulse.

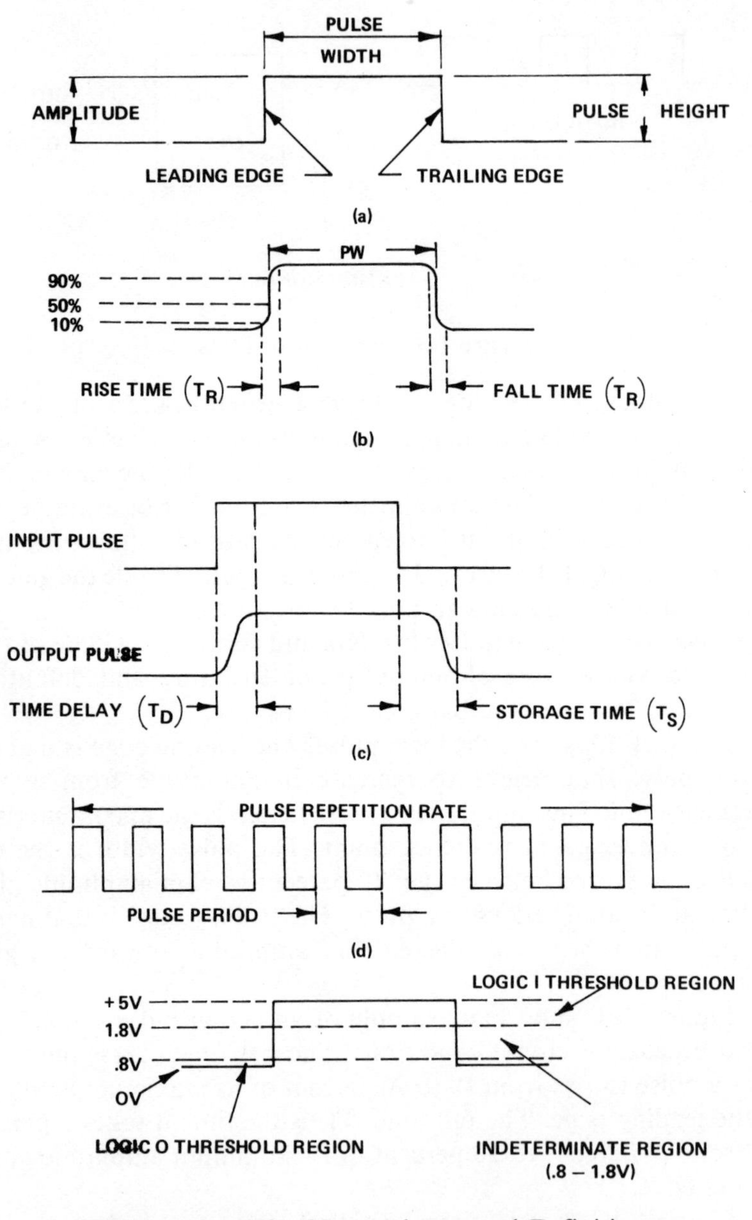

Figure 1-4. Pulse Nomenclature and Definitions (a) Ideal Pulse (b) Actual Pulse (c) Time Delay (d) Pulse Frequency (e) Pulse Threshold Voltages

Figure 1-4D shows that the pulse period is the time the pulse first begins to increase in amplitude, falls back to the reference level, and then begins to increase again in amplitude. Also, the number of pulses that occur in one second is referred to as the pulse repetition rate (PRR) or frequency. When the 10 pulses shown occur in 1 sec of time, the pulse repetition rate is 10 hertz. If you measured the pulse period and it was 10 μ sec long, you could then find the PRR with the following formula:

$$PRR(F) = \frac{1}{PP(\text{pulse period})} = \frac{1}{10 \times 10^{-6}\ \text{sec}} =$$

$$.1 \times 10^{+6}\ \text{hertz} = 100000\ \text{hertz} = 100\text{K hertz}$$

Similarly, you could find the pulse period by using the formula:

$$PP = \frac{1}{PRR} = \frac{1}{1 \times 10^{-5}\ \text{sec}} =$$

$$.00001\ \text{sec} = 10\ \mu\ \text{sec}$$

If the pulse period was symmetrical (the high and low states were equal in time, then the actual pulse width would be 5 μ sec.

Figure 1-4E shows the amplitude threshold regions of a pulse for the proper operation of a digital circuit. The 0 bit threshold region is between 0v and +.8v. The 1 bit threshold region is between 1.8v and +5v. The indeterminate region between .8v and 1.8v means that the digital circuit may or may not respond to the desired condition that the circuit was intended for.

Positive and Negative Logic

So far we have mentioned the 1 bit as being at +5 volts and the 0 bit being at 0 volts. This is called positive logic where the 1 bit is represented by the more positive voltage. Figure 1-5A also shows another representation of positive logic where 0 volts is equal to the 1 bit and −5v is equal to the 0 bit. Figure 1-5B shows the voltage levels representing negative logic where the 1 bit is the more negative voltage. In the first case, the 1 bit is equal to 0 volts and the 0 bit is equal to +5 volts. In the second case, the 1 bit is equal to −5 volts and the 0 bit is equal to 0 volts.

The design engineer usually tries to use one or the other type of logic, since it simplifies both design and servicing of digital equipment. There are, however, times when both positive and negative logic are used in the same piece of equipment. The 7400 TTL-IC digital circuits referred to and used in this book utilize positive logic.

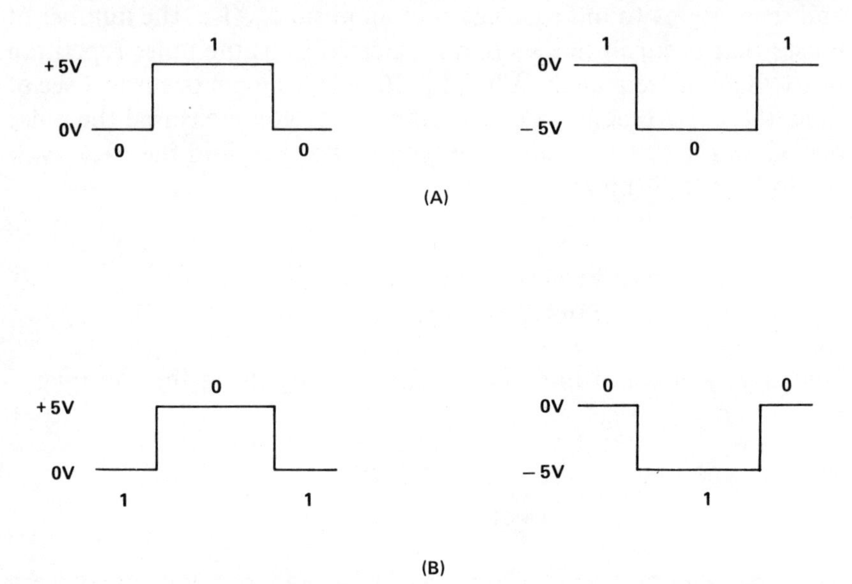

Figure 1-5. Voltage Representation of Bits
(a) Positive Logic (b) Negative Logic

1-2 CONVERTING FROM DECIMAL TO BINARY

As you know the numbers in the decimal system are 0,1,2,3,4,5,6,7,8,9 and the base or radix of the system is 10. The numbers in the binary system are 0 and 1, and the base of the system is 2. Therefore, the decimal numbers must be represented as binary numbers using only 0 and 1. The following chart shows the decimal numbers with their binary equivalents.

Decimal	Binary
0	0
1	1
2	10
3	11
4	100
5	101
6	110
8	1000
9	1001

It would be helpful to you in using binary numbers to memorize this chart.

To convert a decimal number to a binary number, you divide the decimal number by 2 and the remainder from each division results in the binary number.

Example: Convert 25_{10} to binary;

divide by 2)25
divide by 2)12 with 1 remainder
divide by 2) 6 with 0 remainder
divide by 2) 3 with 0 remainder
divide by 2) 1 with 1 remainder
0 with 1 remainder

The binary number appears as the remainder and is written from the bottom up, 11001_2.

$$\text{Thus: } 25_{10} = 11001_2$$

The subscript number indicates the base of that particular number.

Two more examples of converting from decimal to binary are given:

Examples: $13_{10} = \underline{\quad ? \quad}_2$

2)13
2) 6 1
2) 3 0
2) 1 1
0 1

$13_{10} = \underline{1101}_2$

$72_{10} = \underline{\quad ? \quad}_2$

2)72
2)36 0
2)18 0
2) 9 0
2) 4 1
2) 2 0
2) 1 0
0 1

$72_{10} = \underline{1001000}_2$

1-3 CONVERTING FROM BINARY TO DECIMAL

The decimal number system can be expanded by the use of powers, (exponents):

$$1 \times 10^0 = 1$$
$$1 \times 10^1 = 10$$
$$1 \times 10^2 = 100$$
$$1 \times 10^3 = 1000$$

and so on . . . . .

For example the number $5{,}241_{10}$ can be written:

$$
\begin{array}{l}
5{,}241 \\
\rightarrow 1 \times 10^0 = \quad 1 \\
\rightarrow 4 \times 10^1 = \quad +40 \\
\rightarrow 2 \times 10^2 = \quad +200 \\
\rightarrow 5 \times 10^3 = \quad \underline{+5000} \\
\qquad\qquad\qquad\quad 5{,}241
\end{array}
$$

Notice that as the digit place moves to the left of the decimal point, the power or exponent increases.

$$
\begin{array}{cccc}
(10^3) & (10^2) & (10^1) & (10^0) \\
5 & 3 & 4 & 1_{10}
\end{array}
$$

The binary system is also expanded by the use of powers:

$$1 \times 2^0 = 1$$
$$1 \times 2^1 = 2$$
$$1 \times 2^2 = 4$$
$$1 \times 2^3 = 8$$

and so on . . . . .

For example the number 101_2 can be written:

$$
\begin{array}{l}
101 \\
\rightarrow 1 \times 2^0 = \quad 1 \\
\rightarrow 0 \times 2^1 = \quad +0 \\
\rightarrow 1 \times 2^2 = \quad \underline{+4} \\
\qquad\qquad\qquad 5
\end{array}
$$

Therefore, $101_2 = 5_{10}$

Here again, as the bit place moves to the left of the binary point, the power increases.

$$
\begin{array}{ccc}
(2^2) & (2^1) & (2^0) \\
1 & 0 & 1
\end{array}
$$

This is known as positional notation and if you raise (multiply) each 2 to its exponent you could place the raised power or products directly over each bit as shown.

$$\begin{array}{ccc} 4 & 2 & 1 \\ 1 & 0 & 1 \end{array}$$

Now, converting from binary to decimal becomes even more easy. You just add up the raised powers of the positions with a 1 to find the decimal equivalent.

$$\begin{array}{ccc} 4 & 2 & 1 \\ 1 & 0 & 1_2 \end{array} = 4 + 1 = 5_{10}$$

For another example, let's convert the binary number 1101101 to its decimal equivalent.

First place the raised powers over each bit position.

$$\begin{array}{ccccccccc} & 64 & 32 & 16 & 8 & 4 & 2 & 1 & \\ \text{MSB} \longrightarrow & 1 & 1 & 0 & 1 & 1 & 0 & 1 & \longleftarrow \text{LSB} \end{array}$$

Notice that beginning with the 1 on the extreme right which is termed the least significant bit (LSB), you place a 1 directly above it. Then moving left to the next highest order bit, you double the 1 and place 2 above that position. Again, you move left to the next highest order bit and double the 2 and place 4 above that position. You continue this procedure until the maximum significant bit (MSB) on the extreme left is complete.

To finish the conversion, all you have to do is add up the raised powers for each position that contains a 1:

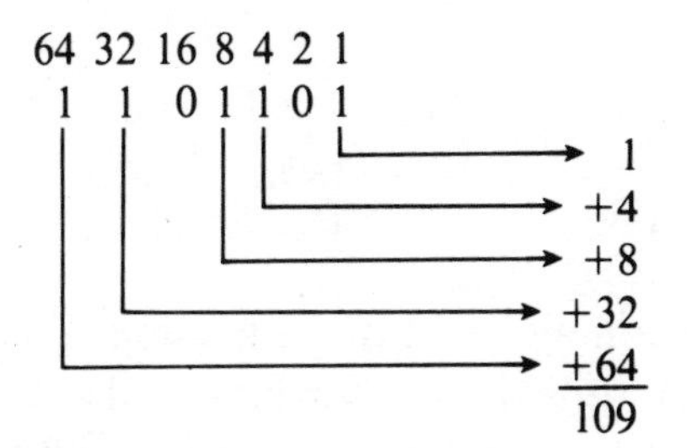

Therefore: $1101101_2 = 109_{10}$

Two more examples of converting from binary to decimal are given:

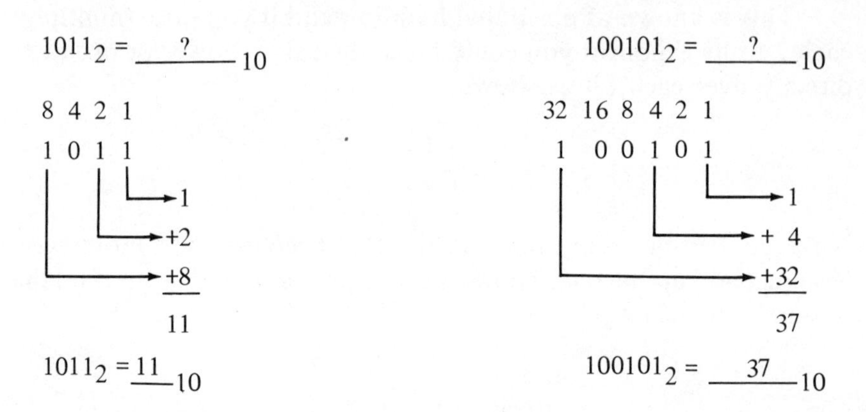

The quickest way to learn conversion from decimal to binary and binary to decimal while double-checking your answers, is to take a paper and pencil and arbitrarily write down some decimal numbers. Now convert these numbers to binary. Take these binary numbers and convert them back to the decimal numbers. If the decimal numbers are the same, you've got it!

1-4 USING BINARY ARITHMETIC

The binary arithmetic operations of addition, subtraction, multiplication and division are essentially the same as for decimal arithmetic operations. There are some slight differences that will be covered as they are encountered.

Binary Addition

When adding with binary numbers there are only four conditions that will be met:

0	0	1	1
+0	+1	+0	+1
0	1	1	10

The last condition may be a little confusing. You know that 1 + 1 = 2 in decimal. The same is true for binary also because $2_{10} = 10_2$. We add 1 + 1 in binary by saying, "1 plus 1 equals 0 with a carry of 1."

When adding larger binary numbers we carry the resulting 1's into higher-order columns as shown on the next page:

	Binary	Decimal Equivalent
	1 ← carry	
(augend)	1 0 1	5
(addend)	+ 0 0 1	+ 1
(sum)	1 1 0	6

As another example:

Binary		Decimal Equivalent
1 1 ←	carry →	1
1 0 0 1		9
+ 0 0 1 1		+ 3
1 1 0 0		12

In the next example, a 1 carry results in three 1's being added together in a column.

Binary		Decimal Equivalent
1 1 1	carry →	1
1 1 1		7
+ 0 1 1		+ 3
1 0 1 0		10

In column two, we just subtotal (add) the carry from the first column and the 1 in the augend. This results in 0 with a carry of 1, which we place in the third column. We now continue to add the second column of 0 + 1 in the addend equals 1. We add column three normally with its resulting carry taken to the next highest-order position. You can use this subtotal method regardless of the number of 1's in a single column. Just be sure to place each 1 carry into the next high-order position or column.

Try some of your own binary additions and check them by using binary to decimal conversion.

Binary Subtraction

When subtracting binary numbers, there are only four conditions that will be met:

0	1	1	0	
−0	−0	−1	−1	
0	1	0	1	(with a borrow)

The last condition is probably the most confusing in binary arithmetic. Obviously you cannot subtract 1 from 0 unless there is a 1 in the next highest-order position to borrow from. An example of this is shown below:

	Binary		Decimal Equivalent
	0	borrow	
(minuend)	1 ~~1~~ ¹0		6
(subtrahend)	− 0 0 1		−1
(difference)	1 0 1		5

When you borrow a 1 from column two and place it in column one, you have to think as if this column contained a 2, since $2_{10} = 10_2$. Of course $2_{10} - 1_{10} = 1_{10}$ and $10_2 - 1_2 = 1_2$.

The next example will help clarify this by placing parentheses around the borrowed numbers.

Binary	Decimal Equivalent
1	
⁰~~1~~ (~~¹0~~) (¹0) 1	9
− 0 0 1 1	−3
1 1 0	6

Notice that the borrowed number must be moved two columns to the right. We have to borrow a 1 from column three containing (10), leaving a 1 in column three and making column two (10). We now can proceed with the normal subtraction.

Again, try some of your own binary subtractions and check them by using binary to decimal conversion.

Subtraction by Complementing

Another method of subtracting which uses less digital circuitry is by complementing and adding. There are two forms to this method of subtraction. The first form is called the 1's complement. The example below will show how this is accomplished.

```
Normal Subtraction          1's Complement

   1 0 1               1 0 1          1 0 1
  -0 1 1             - 0 1 1 ---->  + 1 0 0
  ------             --------       --------
   0 1 0                            1 0 0 1
                                    |---> +1
                                    --------
                                      0 1 0
```

The number in the subtrahend 011 is complemented by changing each 1 to a 0 and each 0 to a 1. This complemented number 100 is then added to the minuend 101. The last 1 carry is brought around and is added to produce the difference of 010 and is referred to as "end-around carry." Another example shown below should help to clarify this.

```
Normal Subtraction                     1's Complement

   1 0 1 0             1 0 1 0            1 0 1 0
 - 0 1 0 1           - 0 1 0 1 ---->    + 1 0 1 0
 ---------           ---------          ---------
     1 0 1                              1 0 1 0 0
                                        |------> +1
                                        ---------
                                          0 1 0 1
```

The second form of complementing is called the 2's complement. The example below will show how this is accomplished.

```
Normal subtraction                    2's complement

   1 0 1             1 0 1                         1 0 1
 - 0 1 1            -0 1 1 ---->100 + 1  =       + 1 0 1
 -------            ------                       -------
   0 1 0                                         1 0 1 0
                                                 |
                                                 |   drop last carry
                                                 v
```

The subtrahend (011) is complemented as before (100) but now an extra 1 is added to it. This 2's complemented number 101 is then added to the minuend 101. The last carry is omitted or dropped and the difference is 010. Shown below is another example of 2's complement subtraction.

```
Normal subtraction              2's complement

  1 0 1 0            1 0 1 0                          1 0 1 0
- 0 1 0 1          - 0 1 0 1 → 1 0 1 0 + 1 =          1 0 1 1
---------          ---------                        ---------
  0 1 0 1                                           1 0 1 0 1
                                                    |
                                                    ↓ drop last carry
```

Binary Multiplication

Binary multiplication is performed the same way as decimal multiplication by shifting and adding, but is much easier since you need only multiply by 1 or 0 as shown in the following examples.

```
    1 0 1          1 0 0
  x 0 1 1        x 1 0 1
  -------        -------
    1 0 1          1 0 0
  1 0 1          0 0 0
  -------      1 0 0
  1 1 1 1      ---------
               1 0 1 0 0
```

Binary Division

Binary division is also performed the same way as decimal division by comparing, shifting and subtracting and is also easier because of using 1 or 0 as shown in the following examples.

```
         1 0 1                 1 0 1
 1 1 ) 1 1 1 1     1 0 0 ) 1 0 1 0 0
       1 1                 1 0 0
       ---                 -----
         1 1                 1 0 0
         1 1                 1 0 0
         ---                 -----
         0 0                 0 0 0
```

1-5 WHY OTHER NUMBER SYSTEMS ARE USED

Other number systems are used to make it easier and more efficient to communicate with digital circuits and computers. These number systems are multiples of the binary system.

Octal Number System

The octal number system has a base of 8 and its numbers are 0,1,2,3,4,5,6,7. If you are counting up in octal, after reaching 7, you would then begin with 10 because the arithmetic symbols 8 and 9 are nonexistent. The same is also true of the following: 20 follows 17, 30 follows 27, 40 follows 37, 50 follows 47, 60 follows 57, 70 follows 67, 100 follows 77, ... etc.

Since $2^3 = 8$, you can see that the octal system is a multiple of the binary system and conversion between the two is relatively easy. An example of this is to convert 27_8 into binary.

$$27_8 = 010\ 111_2$$

All you have to do is to write each octal digit in binary, using 3 bit places. Remember to keep them in the same order. Converting from binary to octal is just as simple as shown:

$$110011101_2 = \underline{\quad ? \quad}_8$$
$$110\ 011\ 101_2 = 635_8$$

Group the bits into groups of three going from right to left and convert each group to its octal equivalent.

When converting from decimal to octal, use the same procedure as with the binary method, except divide by 8. The remainder will be the octal number. When converting from octal to decimal, use the positional notation method as with the binary method, except use the base 8 number system i.e., $8^0 = 1$, $8^1 = 8$, $8^2 = 64$, etc.

Hexadecimal Number System

The hexadecimal number system has a base of 16. Since there is no singular arithmetic symbol above 9, the alphabetic symbols A through F are also used. The numbers in the hexadecimal system are 0,1,2,3,4,5,6,7,8,9,A,B,C,D,E,F.

Since $2^4 = 16$, the hexadecimal system is also a multiple of the binary system. To convert from binary to hexadecimal you simply group the bits into groups of four and then convert each group. As an example:

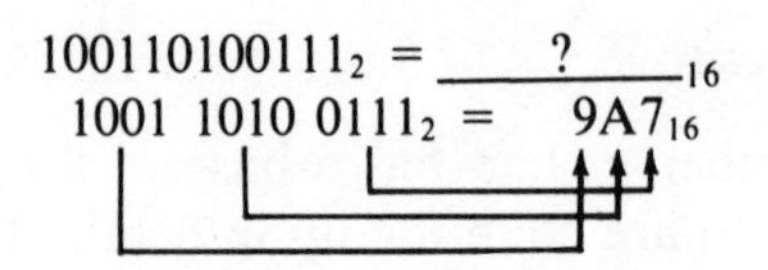

$100110100111_2 = \underline{\quad ? \quad}_{16}$

$1001\ 1010\ 0111_2 = 9A7_{16}$

The conversion from hexadecimal to binary is also similar as shown,

$6\ B\ 5_{16} = 0110\ 1011\ 0101_2$

Converting from decimal to hexadecimal and vice versa, uses the same procedure as for binary and octal, except base 16 is used.

Binary Coded Decimal

Binary Coded Decimal known as the BCD8421 code or, simply BCD code, uses four bit places to represent each of the ten decimal digits. BCD is not a true number system with expanding powers, but rather a code to make it easier to convert between decimal and binary. The following example will help to demonstrate this.

$572_{10} = \underline{\quad ? \quad}_{BCD}$

$572 = 0101\ 0111\ 0010_{BCD}$

Likewise:

$100100100110\ BCD = \underline{\quad ? \quad}_{10}$

$1001\ 0010\ 0110 = 9\ 2\ 6_{10}$

You will use BCD-to-decimal decoders in Chapter 7.

Table 1-1 shows the comparison of the different number systems.

Decimal	Binary	Octal	Hexadecimal	BCD
0	0	0	0	0000
1	1	1	1	0001
2	10	2	2	0010
3	11	3	3	0011
4	100	4	4	0100
5	101	5	5	0101
6	110	6	6	0110
7	111	7	7	0111

Table 1-1. Comparison of Number Systems

Decimal	Binary	Octal	Hexadecimal	BCD
8	1000	10	8	1000
9	1001	11	9	1001
10	1010	12	A	00010000
11	1011	13	B	00010001
12	1100	14	C	00010010
13	1101	15	D	00010011
14	1110	16	E	00010100
15	1111	17	F	00010101

Table 1-1. (continued)

SUMMARY

Digital circuits operate with pulses. The binary number system with a base of 2 is ideal for digital circuits, where a pulse represents 1 and the absence of a pulse represents 0. The voltage levels assigned to 0 and 1 determine if the type of logic used is positive or negative.

Binary arithmetic is very similar to decimal arithmetic, but you must remember how the carry in addition and the borrow in subtraction are treated. The use of 1's and 2's complement makes subtraction easier and requires less components for digital circuits.

Other number systems such as octal with a base 8, hexadecimal with a base 16 and BCD, which is a special binary code, make manipulations with numbers from decimal to binary easier and more efficient in digital circuits and computers.

1-6 PROJECT #1: BUILDING A 5-VOLT REGULATED POWER SUPPLY

The 5-volt regulated power supply is part of the digital IC breadboard/tester. The base of the breadboard/tester is a 7-x 10-inch perforated board, which henceforth will be called the perfboard. The perfboard can be mounted on an angled chassis made of wood or aluminum as shown in Figure 1-6. Angling the breadboard/tester in such a manner makes it easier to wire the IC projects.

A list of recommended tools is given which will aid greatly in the construction of the projects in this book.

Midget long-nose pliers
Midget diagonal cutters
Wire strippers

⅛-inch blade screwdriver
¼-inch blade screwdriver
Standard pliers
37½-50-watt soldering iron
Soldering aid
Set of small files
Electric hand drill
Assorted drill bits up to ½-inch in diameter
Electric jig saw or hack saw

Miscellaneous parts also needed for the projects are: 60/40 (tin-nickel) alloy .030-.040-inch diameter, resin-core solder, assorted machine screws, serrated washers, nuts, terminal solder lugs, solder-type terminal strips, various colored number 20, 22, 24 gauge stranded wire, number 18 and 22 gauge bus wire, and spaghetti. The perfboard for the breadboard/tester has 1/16-inch diameter holes on 15/16-inch centers. Another type of perfboard which will accommodate ICs directly has .04-inch diameter holes on .1-inch centers and will be used in other projects. Various size pieces of aluminum are needed to form the chassis for the projects.

Figure 1-6. Perfboard and Chassis

Soldering Techniques

Even if you are experienced at soldering, the following procedures could serve as a review and also show you some new techniques for breadboarding circuits.

1. Use a low-wattage soldering iron or gun, usually between 37½-50 watts.
2. Tin the tip of the iron according to the manufacturer's instructions and clean the tip between soldering operations by rubbing it on a damp cloth or sponge. This is very important for a fast and clean solder connection.
3. Use a resin core solder because it is non-corrosive to electronic components. A 60/40 tin/nickel alloy solder with a diameter of .030-.040-inch works well for semiconductor circuits.
4. Clean the parts to be soldered, if needed, with fine sandpaper or solvent.
5. First heat the connection where solder is to be applied for a couple of seconds. Then apply the solder to the connection only.
6. Use as little solder as possible, making sure the connection has a sufficient covering of solder and a good bond.
7. You may want to use clip-on heat sinks or long-nose pliers on the semiconductor leads for their protection. This probably won't be necessary providing your iron is clean and you don't leave the iron on the connection for more than a brief period.

You may want to use printed circuit boards. Etched-circuit kits are available at electronic stores. These circuits are sturdier and more reliable than perfboard-constructed circuits, but require more layout work, take longer to construct and are less easy to modify.

With perfboard construction of integrated circuits it is often necessary and faster to use tack-soldering as illustrated in Figure 1-7.

When an electrical connection is needed at the intersection of two wires as shown in Figure 1-7A, tack-soldering may be used as shown in Figure 1-7B. Tack-soldering is also very useful when the connections are close together, as with the IC socket shown in Figure 1-7C. Components may be connected directly with their leads forming a hook splice as shown in Figure 1-7D. A component lead can be used to form a terminal connection by threading it back and forth through the perfboard as shown in Figure 1-7E. The perfboard is a good insulator and cannot be used for the ground return as with a regular aluminum chassis. A ground bus wire as well as a +VCC bus wire can be used with perfboards as shown in Figure 1-7F. Long-nose pliers should be used for bending component and semiconductor leads.

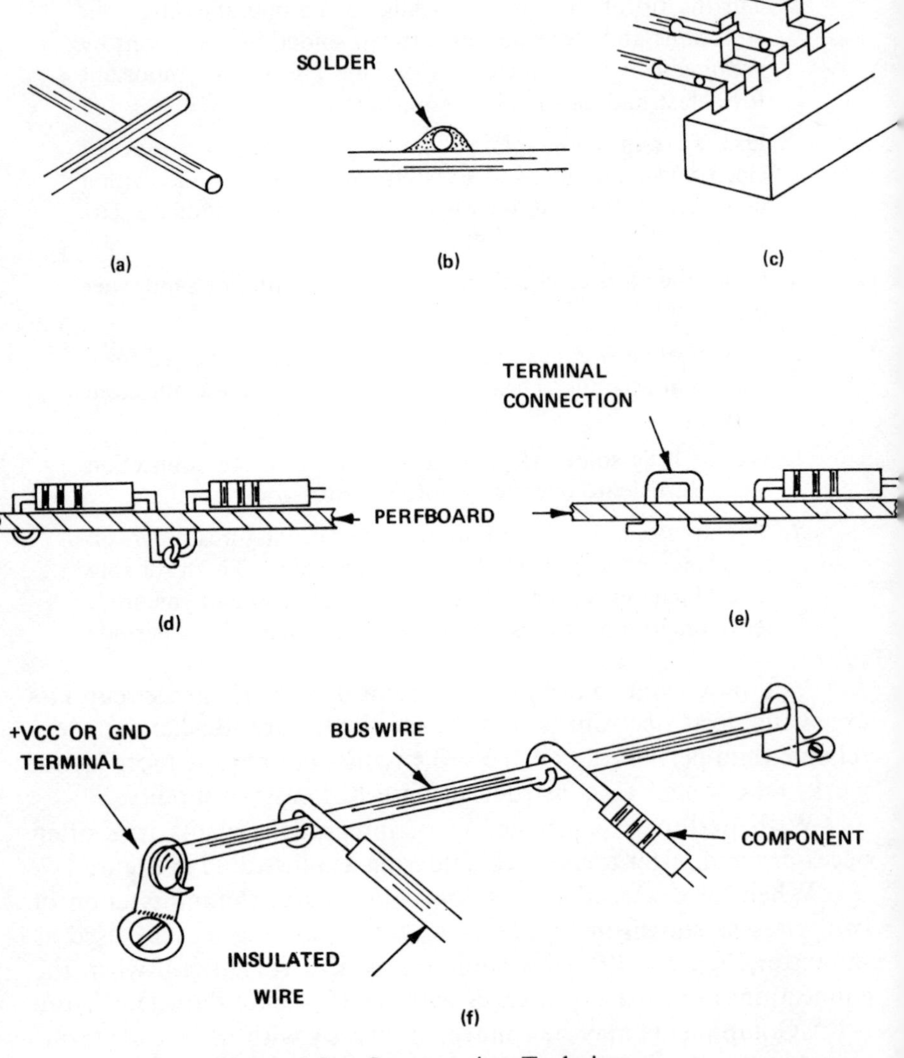

Figure 1-7. Construction Techniques
(a) Wire Intersection (b) Tack-soldering
(c) IC Socket Wiring (d) Hook Splice
(e) Wire terminal (f) Bus wire

The schematic diagram for the power supply is shown in Figure 1-8.

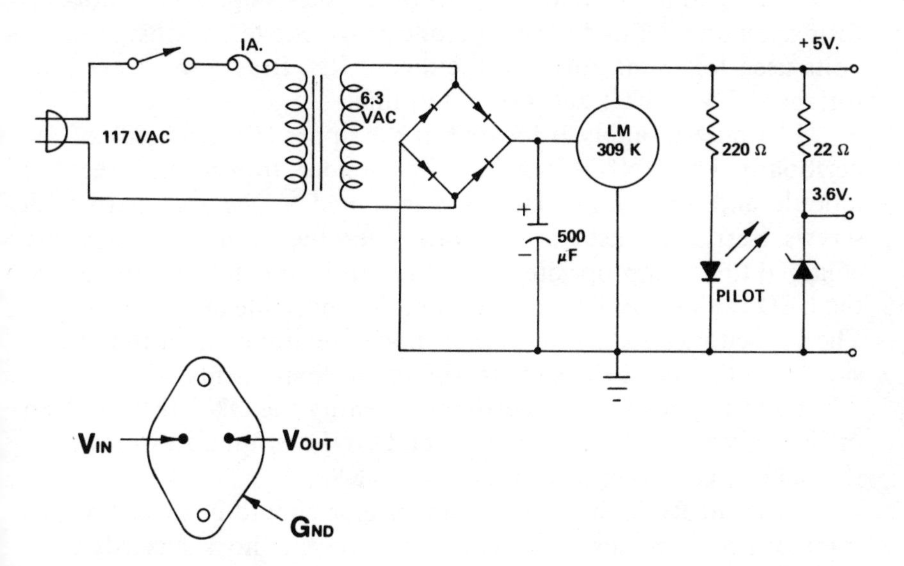

LM 309K BOTTOM VIEW

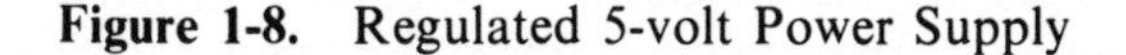

Figure 1-8. Regulated 5-volt Power Supply

Parts List for Project #1

1—AC line cord
1—SPST on/off toggle switch
1—Fuse holder
1—1 amp fuse
1—6.3 volts filament transformer
1—Bridge rectifier module @ 1 amp minimum
1—500 μF@ 50V electrolytic capacitor
1—LM 309K voltage regulator
1—220 ohm ½ watt carbon resistor
1— 22 ohm ½ watt carbon resistor
1—1.6-2.3 volts LED
1—3.6 volts @ 1 watt Zener diode

A 12-volt transformer may be used in place of the 6-volt transformer. The 500 μF capacitor works satisfactorily, but the value can be increased up to 4000 μF, if desired. The LM 309K is a special +5 volt semiconductor voltage regulator. The DC voltage input to this regulator can range from 5 to 35 volts, while the output will remain regulated at +5 volts. The cathode of the LED can usually be identified by the flat side of its base. (See Chapter 8 for more information on LEDs.) The zener diode provides a fairly constant +3.6 volts used for input pulses to the digital ICs. Brass eyelets are used for the +5 V (VCC) and GND terminals.

The power supply is located in the upper left-hand side of the perfboard. The ON/OFF switch, fuse holder, transformer, rectifier module and voltage regulator are mounted to the perfboard with screws, serrated washers and nuts after the mounting holes are enlarged to the appropriate size with a drill. One hole is enlarged so the LED can be pushed through from the underside of the perfboard. The capacitor, resistors and zener diode are mounted on the underside by soldering their leads to the other components.

Labels applied to the perfboard identify the ON/OFF position of the switch, the +5 volts (VCC) and GND terminals as shown on the completed power supply in Figure 1-9.

Caution: Resin epoxy or a rubber glue should be placed on the transformer terminals as insulation to prevent shock hazards from the 117 VAC line.

The projects in this book are intended to serve as a guide. Your actual construction techniques and finished projects will depend upon the materials that are available and your own creativeness.

Figure 1-9. Completed Power Supply

CHAPTER 2

Understanding Digital Building Blocks—Basic Logic Gates

Basic logic gates are the digital building blocks from which virtually all other digital circuits and systems can be built. Miniaturization and the use of micro-electronic technology have reduced complex circuits to the same physical size of these basic logic gates as you will see in later chapters of this book.

This chapter will provide you with the necessary knowledge to trace through complex logic diagrams and understand the function of other related digital devices.

Wiring and testing each of these basic logic gates on the IC breadboard as shown in **Project # 2,** will enable you to get the feel of digital integrated circuits and develop the necessary skills needed for wiring and servicing these types of electronic devices.

Digital circuits are nothing more than switching circuits. In order to electronically accomplish a certain task, a flow of pulses must have a logical plan. The switching of these pulses is performed by logic gates. Logic gates are unlike switches in that they allow data to pass in only one direction, i.e., from input to output, when properly conditioned. Logic gate symbols are functional block diagrams which are much easier for the technician to follow than discrete electronic components. Most manufacturers use standardized logic symbols and this book uses the military-standard-806 (MIL-STD-806) introduced by the U.S. government. This book refers to the 7400 series TTL integrated circuits which adhere to the MIL-STD-806 very closely.

2-1 AND GATE

The symbol and illustration of operation for the AND gate are shown in Figure 2-1.

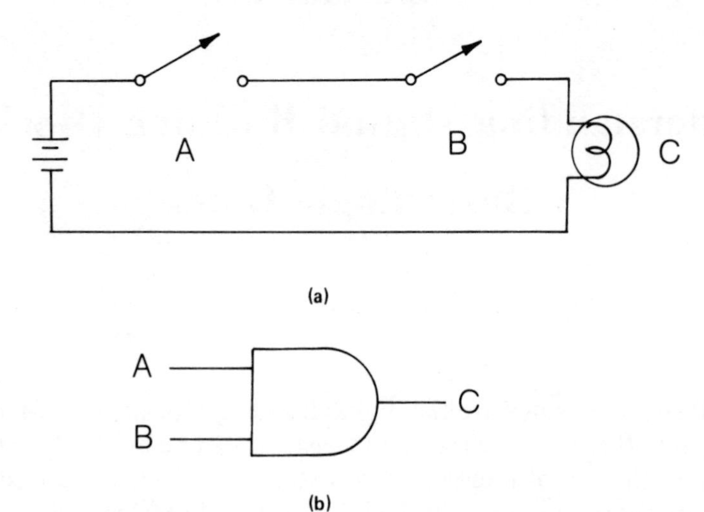

$$A \cdot B = C$$

(c)

INPUTS		OUTPUT
A	B	C
0	0	0
0	1	0
1	0	0
1	1	1

(d)

Figure 2-1. The AND Gate
(a) Electrical Analogy (b) Logic Symbol
(c) Formula (d) Truth Table

When all inputs to an AND gate are 1 at the same time, the output will be 1. The light C will glow, when switch A and switch B are closed. The formula may help to understand the AND gate. It states that "A AND B equals C." The symbol for multiplication (•) stands for AND in Boolean algebra. The chart shows the resulting output condition and is called a truth table.

Figure 2-2 shows actual AND gates in IC form.

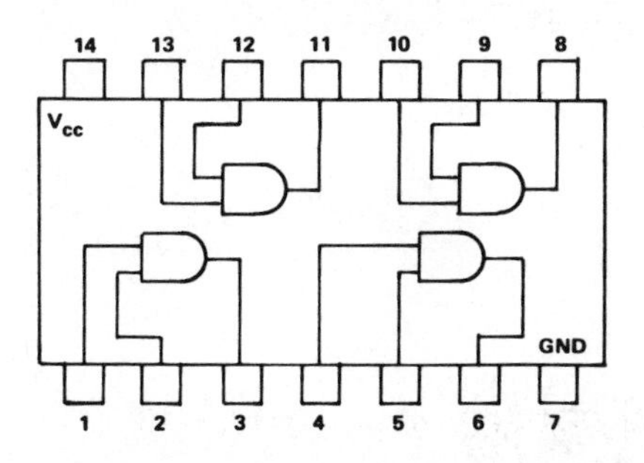

Figure 2-2. IC 7408 Quad-Input AND Gate
(Courtesy Signetics)

Pin 14, marked Vcc, is where the positive 5 volts supply is connected, while pin 7, marked GND, is where the ground is connected. The pins show the AND gates' inputs and outputs respectively.

2-2 OR GATE

The symbol and illustration of operation for the OR gate are shown in Figure 2-3.

When any input to an OR gate is 1, the output will be 1. The light C will glow when switch A is closed or switch B is closed or both switches are closed. The formula for the OR gate states that "A OR B equals C." The symbol for addition (+) stands for OR in Boolean algebra. The truth table shows the various conditions at the inputs with the resulting output condition.

Figure 2-4 shows actual OR gates in IC form.

2-3 INVERTER (NOT GATE)

The inverter, sometimes referred to as a NOT gate, is shown in Figure 2-5.

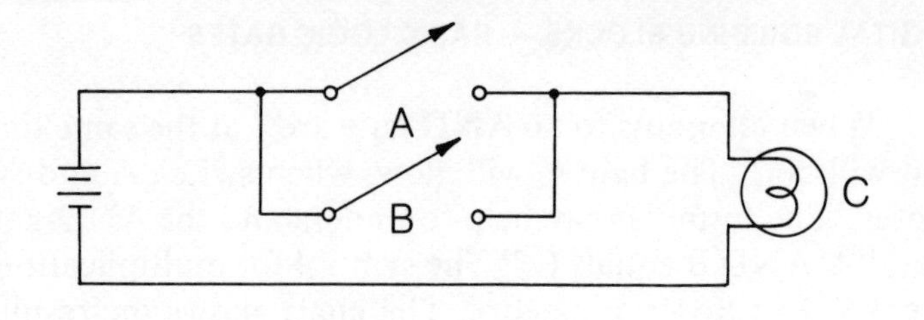

(a)

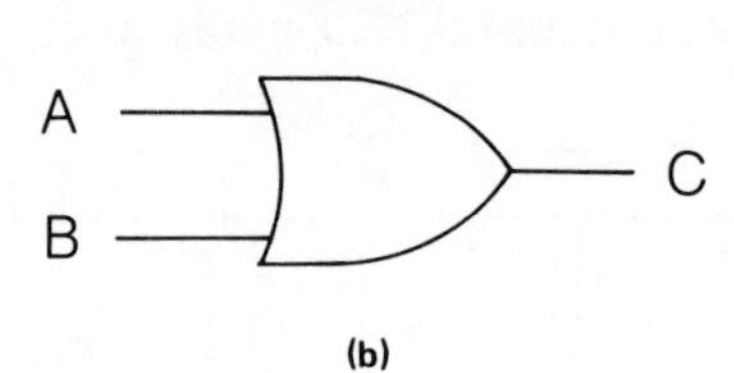

(b)

$$A + B = C$$

(c)

INPUTS		OUTPUT
A	B	C
0	0	0
0	1	1
1	0	1
1	1	1

(d)

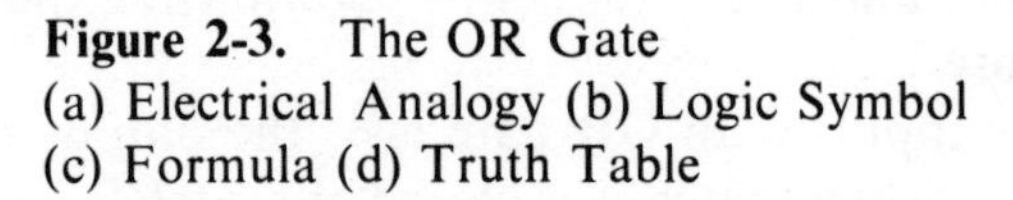

Figure 2-3. The OR Gate
(a) Electrical Analogy (b) Logic Symbol
(c) Formula (d) Truth Table

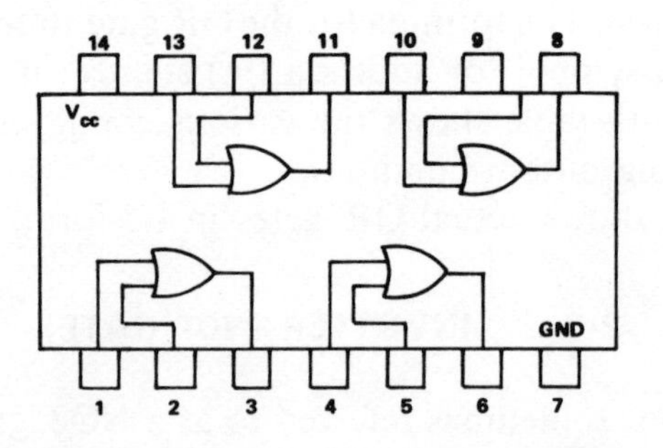

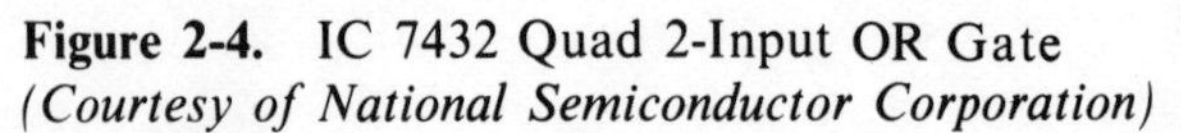

Figure 2-4. IC 7432 Quad 2-Input OR Gate
(Courtesy of National Semiconductor Corporation)

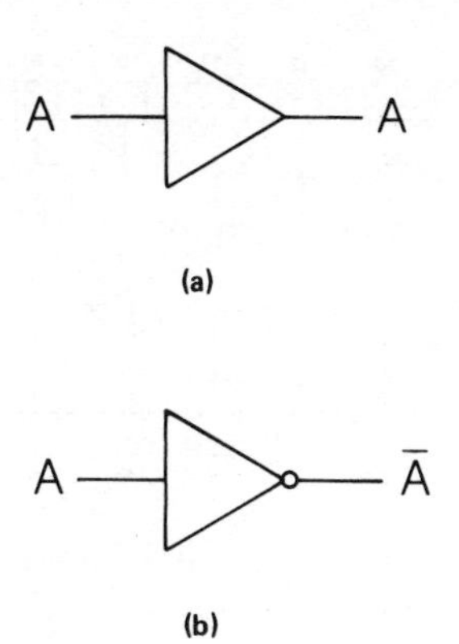

INPUT	OUTPUT
A	$\bar{A}$
0	1
1	0

(c)

Figure 2-5. INVERTER (NOT Gate)
(a) Non-inverter or Driver Logic Symbol
(b) Inverter Logic Symbol (c) Truth Symbol

The non-inverting gate or driver with a single input is shown first. The output of the driver directly follows the input, i.e.; 1 in = 1 out, 0 in = 0 out. This device is used as an amplifier to drive other gates or output indicating devices, or as a buffer between different circuits.

The inverter, also with a single input, has a circle or bubble at the apex of the triangle which indicates negation. When the input is 0, the output will be 1 and when the input is 1, the output will be 0, as the truth table indicates. The bar above the A at the output of the symbol is the Boolean algebra expression for negation and is read "A Not." Some inverters may be shown with the bubble at the input, but the logic function remains the same. Figure 2-6 shows inverters in IC form.

2-4 NAND GATE

The NAND gate, which is one of the most commonly used types of logic gates in digital circuitry, is shown in Figure 2-7.

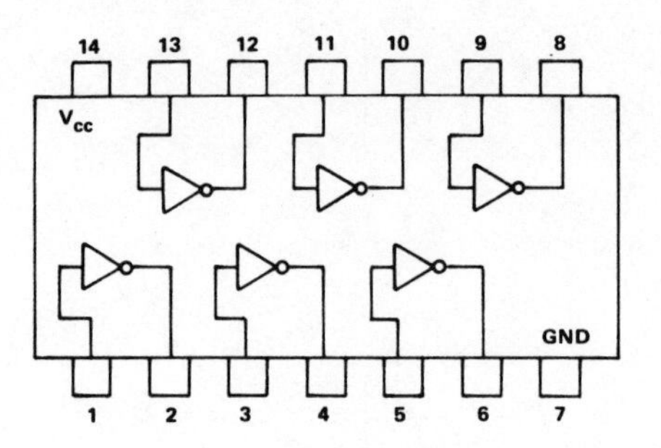

Figure 2-6. IC 7404 Hex Inverter
(Courtesy Signetics)

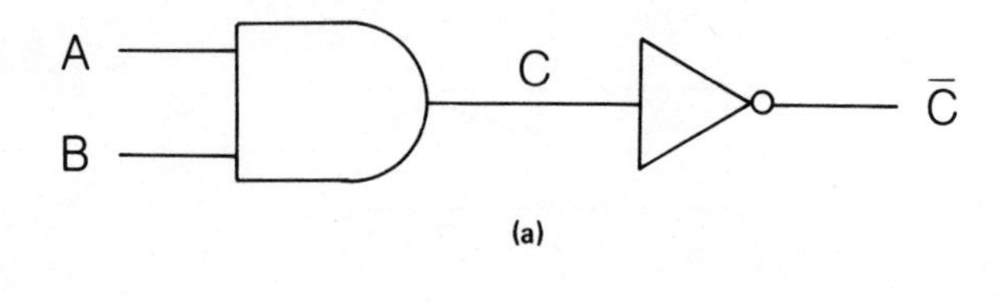

(a)

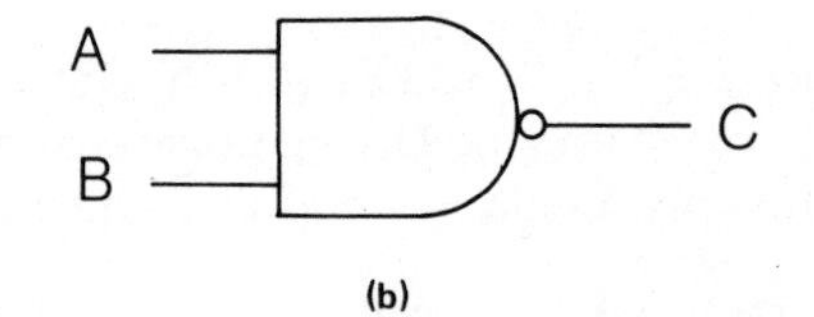

(b)

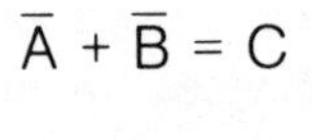

(c)

INPUTS		OUTPUT
A	B	C
0	0	1
0	1	1
1	0	1
1	1	0

(d)

Figure 2-7. The NAND Gate
(a) AND Gate and Inverter (b) Logic Symbol
(c) Formula (d) Truth Table

The NAND gate is actually an AND gate followed by an inverter and may be read "NOT-AND," hence, the term NAND. The standard symbol for the NAND gate omits the triangle of the inverter and uses only the bubble at the output. When any input is 0, the output of the NAND gate will be 1. The formula reads "A not or B not equals C." Inspecting the truth table shows that the NAND gate is a negated input OR gate. Figure 2-8 shows actual NAND gates in IC form.

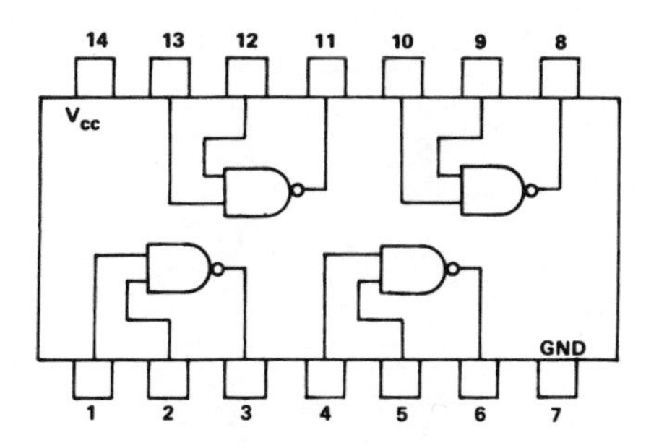

Figure 2-8. IC 7400 Quad 2-Input NAND Gate
(Courtesy Signetics)

If the inputs of a NAND gate are connected together, it then becomes an inverter.

2-5 NOR GATE

The NOR gate, another commonly used logic gate, is shown in Figure 2-9.

The NOR gate is actually an OR gate followed by an inverter and may be read "NOT-OR," hence, the term NOR. When both inputs are 0, the output of the NOR gate will be 1. The formula reads "A not and B not equals C." The truth table reveals that the NOR gate is a negated input AND gate. Figure 2-10 shows NOR gates in IC form.

When the inputs of the NOR gate are connected together, it also becomes an inverter.

2-6 EXCLUSIVE OR GATE

The OR gate previously introduced is termed an inclusive OR gate, because any input or all inputs with a 1 will produce a 1 at the

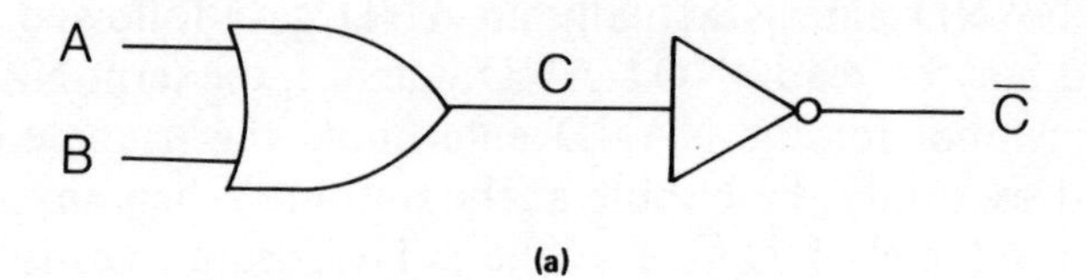

(a)

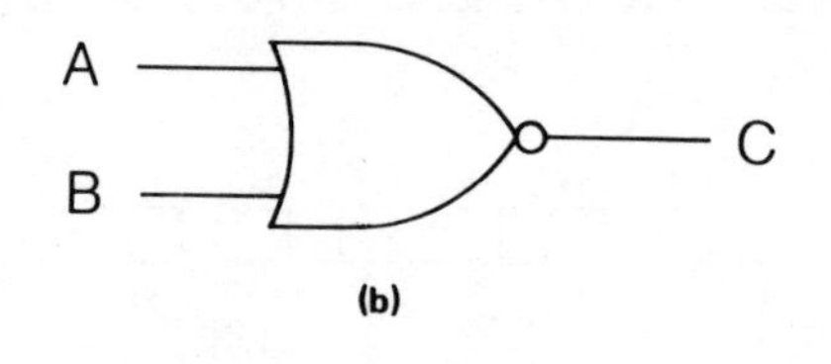

(b)

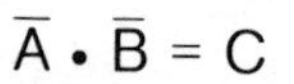

$$\overline{A} \bullet \overline{B} = C$$

(c)

INPUTS		OUTPUT
A	B	C
0	0	1
0	1	0
1	0	0
1	1	0

(d)

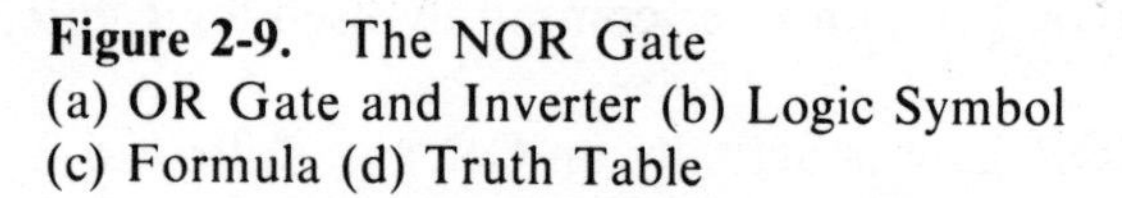

Figure 2-9. The NOR Gate
(a) OR Gate and Inverter (b) Logic Symbol
(c) Formula (d) Truth Table

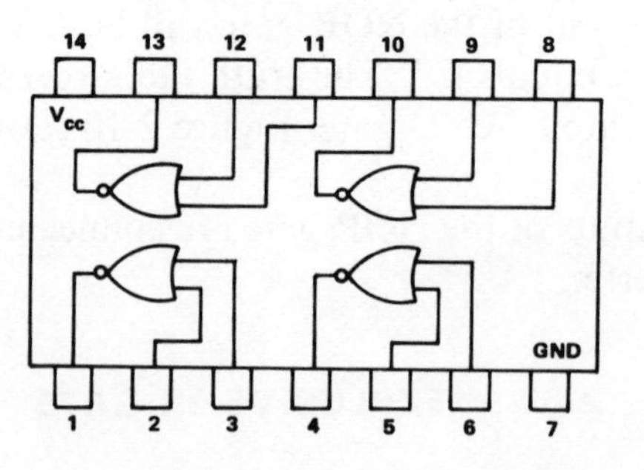

Figure 2-10. IC 7402 Quad 2-Input NOR Gate
(Courtesy Signetics)

output. Some digital circuits require an OR gate to produce a 1 when only one input is a 1 but not any others. This type of OR gate is called an EXCLUSIVE OR gate and is shown in Figure 2-11.

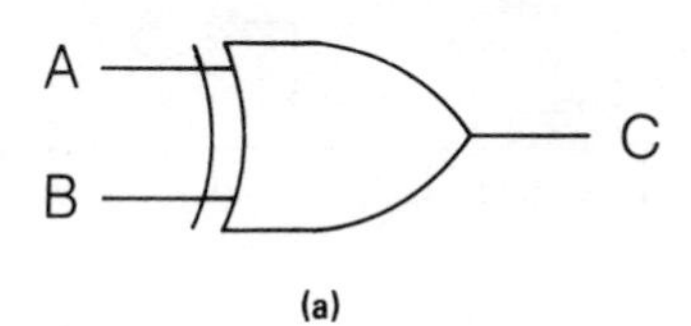

(a)

$$A\bar{B} + \bar{A}B = C$$

(b)

INPUTS		OUTPUT
A	B	C
0	0	0
0	1	1
1	0	1
1	1	0

(c)

Figure 2-11. The EXCLUSIVE OR Gate
(a) Logic Symbol (b) Formula (c) Truth Table

The EXCLUSIVE OR gate symbol is recognized by the extra line drawn across the inputs. The formula reads "A and B Not or A Not and B equals C." The truth table shows that only one input with a 1 will produce an output of 1. Figure 2-12 shows EXCLUSIVE OR gates in IC form.

2-7 NEGATED INPUTS

Some digital circuitry requires that basic gates have some inputs negated in order to achieve a particular function. Figure 2-13 shows an AND gate with one input negated and is sometimes referred to as an inhibit gate.

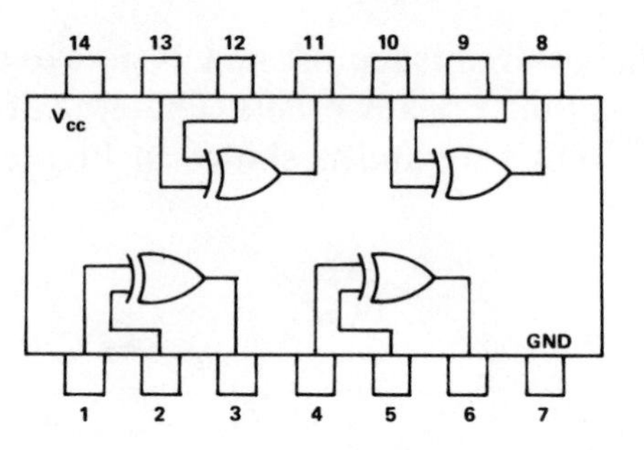

Figure 2-12. IC 7486 Quad 2-Input Exclusive OR gate *(Courtesy Signetics)*

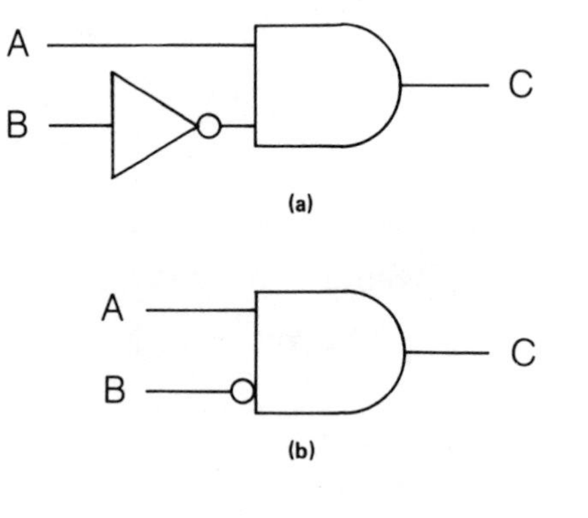

$$A \bullet \overline{B} = C$$

(c)

INPUTS		OUTPUT
A	B	C
0	0	0
0	0	0
1	0	1
1	1	0

(d)

Figure 2-13. The Inhibit Gate
(a) Inverter and AND Gate (b) Logic Symbol
(c) Formula (d) Truth Table

If the gate in Section (b) of Figure 2-13 were needed, but not available, an inverter could be used with an AND gate as shown in section (a). The formula reads "A and B not equals C." The truth table reveals that only one possible input combination will produce a 1 at the output, while the other input combinations inhibit the gating action.

Logic gates that require negated inputs can be replaced by other commonly available gates as shown in Figure 2-14. A negated input OR gate can be replaced with a NAND gate.

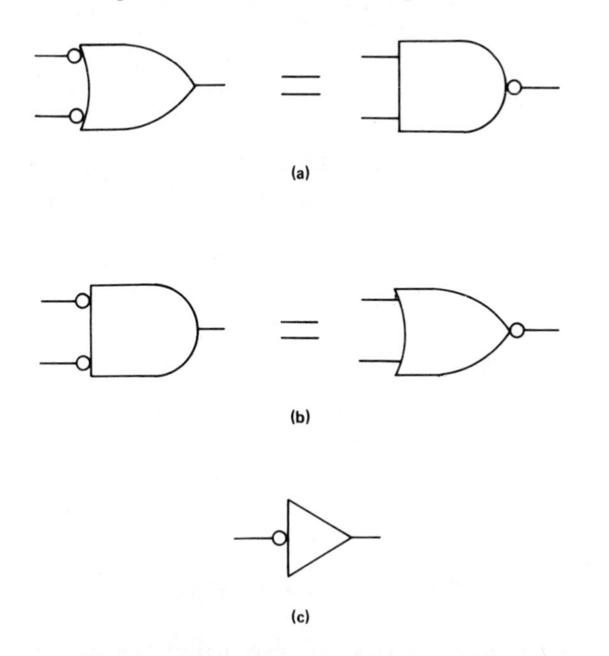

Figure 2-14. Negated Input Gates
(a) Negated Input OR Gate
(b) Negated Input AND Gate
(c) Negated Input Inverter

Likewise, a negated input AND gate can be replaced with a NOR gate. A negated input inverter does not change its function, but may make it easier to understand when reading a logic diagram.

2-8 COMBINATION GATES

Logic gates are combined together to form switching arrangements to perform certain operations or functions. Figure 2-15 illustrates an AND-to-OR gate network.

An output of 1 will be produced if A and B or C and D inputs are 1.

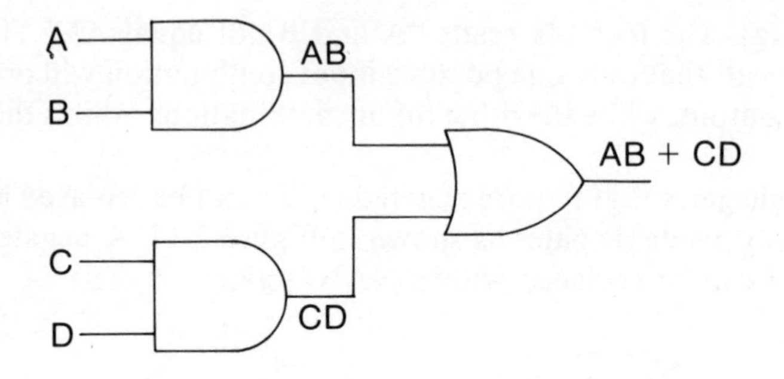

Figure 2-15. AND-to-OR Gate Network

Figure 2-16 illustrates an OR-to-AND gate network.

An output of 1 will be produced if A or B and C or D inputs are 1.

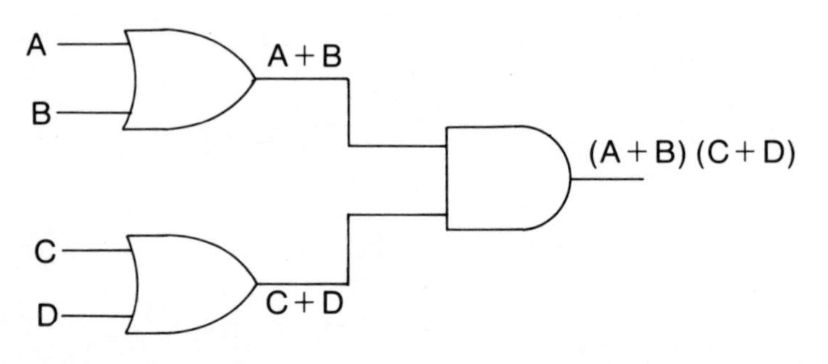

Figure 2-16. OR-to-AND Gate Network

Figure 2-17 shows AND-OR-INVERT (AND-NOR) gates in IC form.

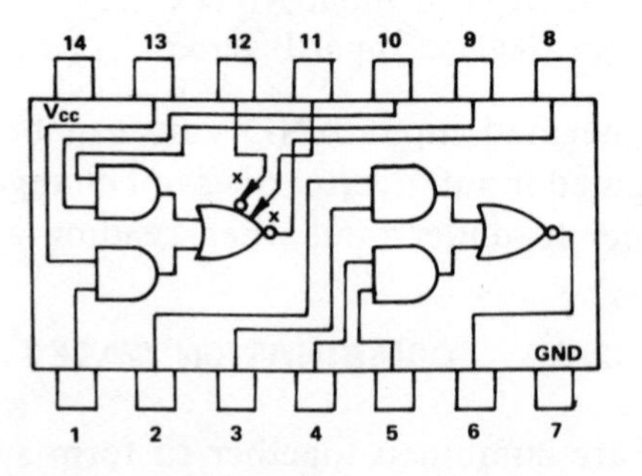

Figure 2-17. IC 7451 Expandable Dual 2-Wide 2-Input AND-OR Invert Gate
(Courtesy Signetics)

An output of 1 will be produced if both of the AND gates feeding the NOR gate fail to produce a 1. The X and $\overline{X}$ lines to the one NOR gate are used with expander ICs to enlarge the logic network.

Different combinations of logic gates are used sometimes to obtain other logic functions. Figure 2-18 illustrates how two INHIBIT gates and one OR gate are combined to produce an EXCLUSIVE OR gate.

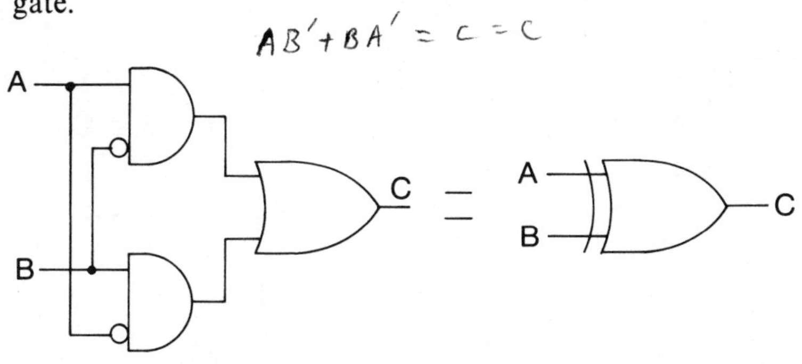

Figure 2-18. Combination of Gates to produce Exlcusive-OR

SUMMARY

The basic logic gates presented in this chapter are the building blocks of all digital logic circuits. The more complex logic functions are, of course, fabricated into ICs with different logic symbols to facilitate the work of the designer and technician. This reduces the number of discrete components used in various electronic equipment.

One of the best ways for you to acquire a working knowlege of these basic gates is to take scrap paper and draw the symbols. Then, applying 0's and 1's to the inputs of your drawing, prove the truth table for each logic gate. Figure 2-19 is a summary of the basic logic gates which is a ready reference that can be used in practicing the operation of the basic logic gates.

2-9 PROJECT #2: BUILDING AN IC BREADBOARD AND TESTING LOGIC GATES

Project #2 continues with the breadboard/tester by constructing the IC sockets, LEDs and logic switches in the right-hand side of the perfboard as shown in Figure 2-20.

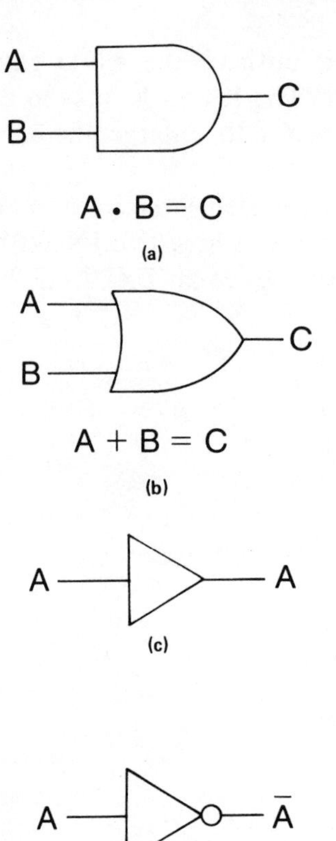

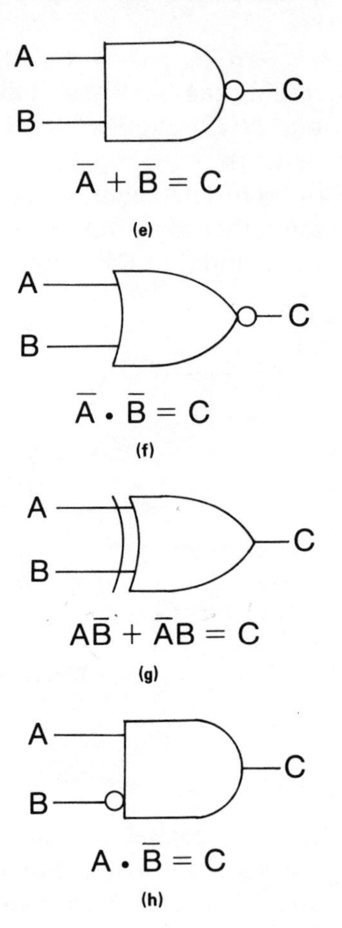

Figure 2-19. Summary of Basic Logic Gates
(a) AND Gate (b) OR Gate (c) Driver
(d) Inverter (e) NAND Gate (f) NOR Gate
(g) Exlusive OR Gate (g) Inhibit Gate

There are a number of terminal-lead methods you can use for breadboarding the various circuit configurations. The method used in this book has taper pins soldered to each end of the wires and uses brass eyelets for their jacks. Other types of methods which might be used are:

Mini-pin plug and jack
Solder terminal and mini-alligator
clip lead

Spring type connections and tinned wires
Mini-banana plug and jack

The terminal-lead method you choose will determine the size perfboard needed to accommodate at least four IC sockets. Therefore, most components should be obtained before layout of the perfboard.

Figure 2-20. Breadboard/Tester (top view)

The most important part of this construction is the initial layout of components. The time spent in this layout work will be saved later when actual construction begins. Lay out the components on the perfboard to get an idea of the approximate distance needed between each one of them. Using a pencil, draw a rough outline of the components and identify them as shown in Figure 2-21. The pencil marks are easily erased later.

There are a variety of IC sockets on the market. Some need to have a hole cut in the perfboard for mounting, while others are bolted directly on the perfboard without needing a hole. The type you use will depend on whether or not you will have to cut the four large rectangular holes in the perfboard. If you have to cut the holes, first enlarge one of the existing holes within the rectangle to accommodate the blade of a jigsaw. Now proceed with cutting the rectangular hole. If the slide switches are used, first drill two holes of the appropriate size within the drawn rectangles. Then with a small file shape the holes so that the slides will move freely when the switches are mounted on the perfboard.

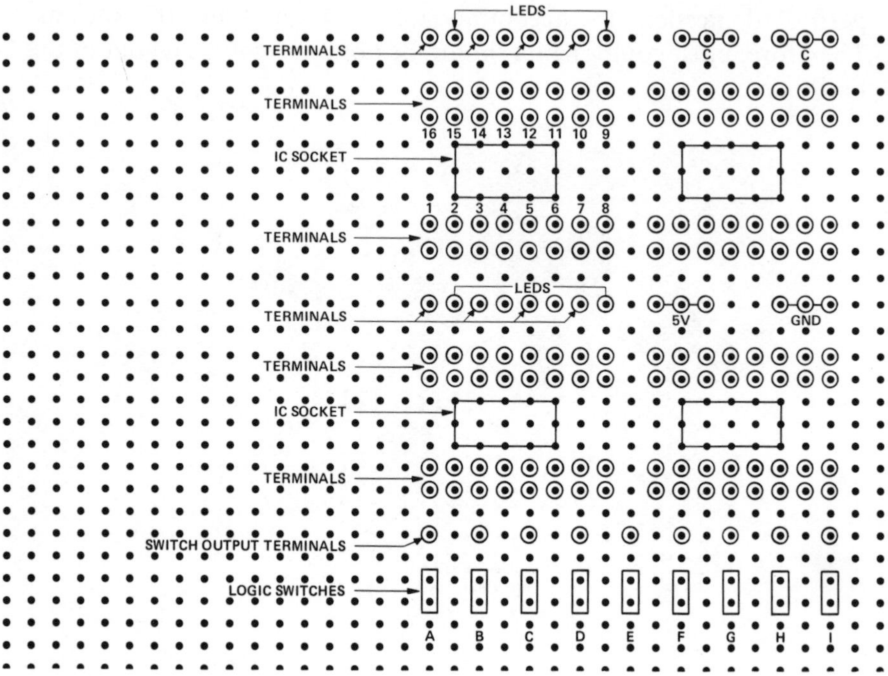

Figure 2-21. Perfboard Layout

The LEDs are pushed through previously enlarged holes from the underside of the perfboard. The LEDs and IC sockets can be glued to the perfboard. Remember to mount all of the IC sockets facing one way. Their direction can usually be determined by an arrow, notch or hole on one end, or some other identifying feature. Notice that each IC socket pin has two terminal connections to facilitate circuit breadboarding. The terminals marked (common), 5v and GND are auxiliary connection points to also make wiring the circuits easier.

After the IC sockets, terminals, LEDs and switches are mounted, the wiring can be completed as shown in Figure 2-22.

A schematic of how the logic switches are wired is shown in Figure 2-23.

The LEDs draw enough current when directly connected to the ICs so that the loading effect can cause erratic operation of other ICs

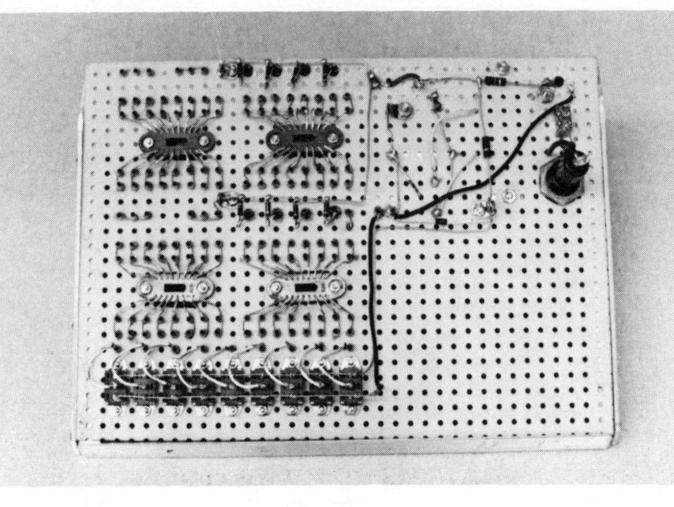

Figure 2-22. Breadboard/Tester (bottom view)

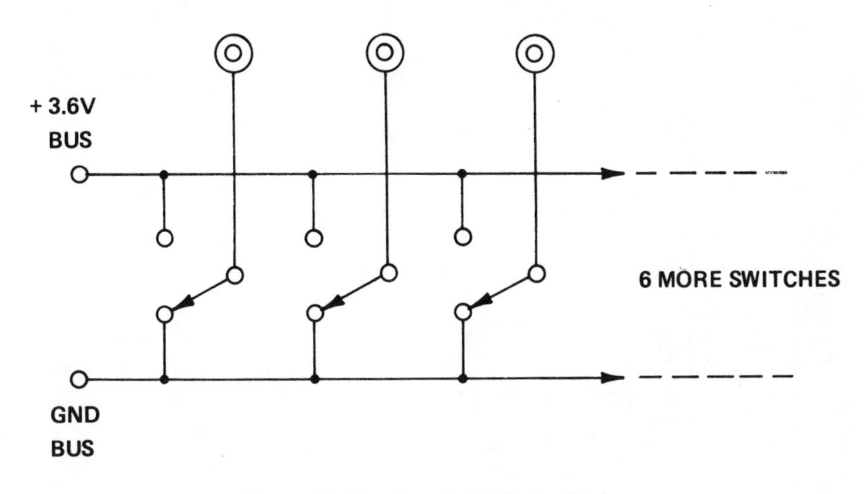

Figure 2-23. Logic Switch Wiring

driven from the same points. Therefore, separate drivers are used for each LED as shown in Figure 2-24.

The dual transistor circuit shown in Figure 2-24A is called a *Darlington Pair*. The 47K Ω base resistor of Q1 decreases the loading effect. When a 1 (positive voltage) appears across this register, Q1 turns on, which also turns on Q2. These transistors now provide a low resistance path for current flow through the LED. The 220 Ω resistor limits the current flowing through the LED to a safe value.

The circuit is constructed in two parts with the transistors and

47K Ω resistor being mounted on a separate smaller perfboard as shown in Figure 2-25.

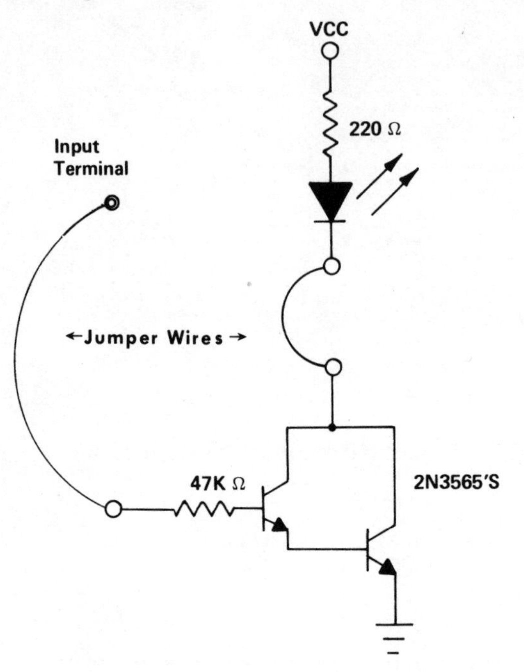

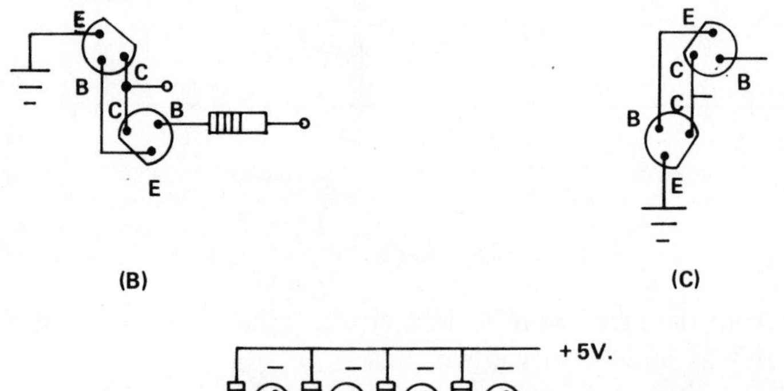

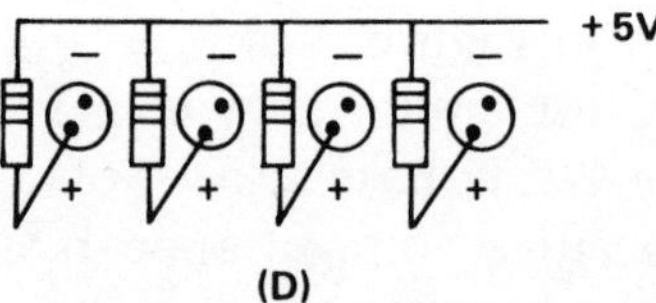

Figure 2-24. LED Drivers (a) Schematic Diagram (b) Top View (c) Bottom View (d) LED—Resistor Bus Line

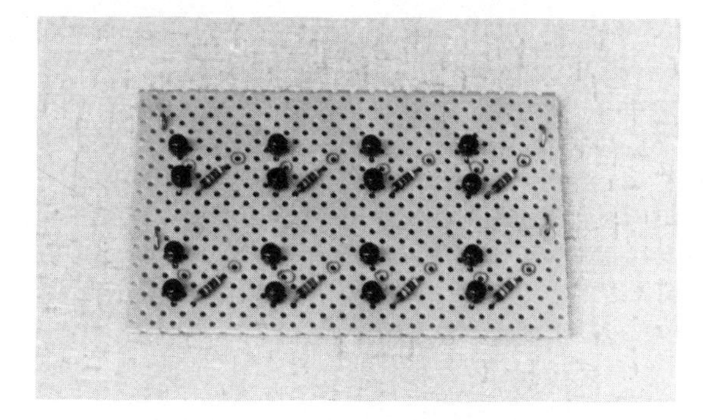

Figure 2-25. LED Driver Perfboard

Figures 2-24B and 2-24C show how the transistors and 47KΩ resistor are wired, while Figure 2-24D shows how the LED and 220 Ω resistor are connected to a +5v bus line on the underside of the breadboard/tester perfboard. The LED driver input terminals are mounted to the left of each LED. The smaller perfboard with the eight individual circuits is mounted to the chassis beneath the breadboard/tester and jumper wires complete the circuits as shown in Figure 2-26.

Figure 2-26. LED Driver Jumper Wire Connections

After the wiring is complete, identifying labels can be attached to the perfboard as shown in Figure 2-20.

Testing Basic Logic Gates

You can now proceed with the testing of the basic logic gate ICs. At this point it will be helpful if you refer to Section 3-5 (Wiring, Soldering and Handling ICs) so as not to damage the ICs. A setup of one of the tests is shown in Figure 2-27.

Figure 2-27. Testing a Logic Gate

Leave the power off until the breadboarding is complete. A 7400 quad 2-input NAND gate has been inserted into the upper left-hand IC socket. A lead wire goes from +5 volts to terminal pin 14 of the IC. A lead wire goes from GND to terminal pin 7 of the IC. A lead wire goes from switch A terminal to terminal pin 1 of the IC. Another lead wire goes from switch B terminal to terminal pin 2 of the IC. The last lead wire goes from terminal pin 3 of the IC to one of the LED input terminals. Power may now be applied to the breadboard/tester.

Place switches A and B in the upper or 1 position. Now, referring to Figure 2-7, prove the truth table. When either or both of the switches are moved to the down or 0 position the LED will light.

Referring to Figure 2-8 connect and test the remaining three NAND gates in a similar manner.

Remember that since we have a 16-pin IC socket, pins 10-16 are actually pins 8-14 respectively when using a 14-pin IC.

You can test all of the basic logic gates presented in this chapter with the same method. Table 2-1 can be used as a guide for each test.

Test # :	Gate Being Tested:	Refer To Figures:	IC's Used:
1.	AND	2-1, 2-2	7408
2.	OR	2-3,2-4	7432
3.	INVERTER	2-5, 2-6	7404
4.	NAND	2-7, 2-8	7400
5.	NOR	2-9, 2-10	7402
6.	EXCLUSIVE-OR	2-11, 2-12 (2-18, 2-2,) (2-4, 2-6)	7486 (7408, 7432) (7404)
7.	INHIBIT	2-13, 2-2, 2-6	7408, 7404
8.	NEGATED INPUT OR	2-14, 2-8, 2-4, 2-6	7432, 7404 7400
9.	AND-TO-OR	2-15, 2-2 2-4	7408, 7432
10.	OR-TO-AND	2-16, 2-2 2-4	7408, 7432
11.	AND-OR-INVERTER	2-17	7451
12.	INVERTERS (NAND, NOR)	2-8, 2-10 2-28	7400, 7402

Table 2-1. Basic Logic Gate Tests

Notice that from test number 6 on, you may use several ICs to perform the logic functions. Remember to wire the VCC and GND pins of each IC. Test number 12 is accomplished simply by connecting the inputs to each gate together as shown in Figure 2-28.

Figure 2-28. Inverters
(a) NAND Inverter (b) NOR Inverter

Parts List for Project # 2

4—16-pin DIP IC sockets
160—eyelets or other type of terminals
9—SPDT mini-switches (slide or toggle type)
8—LEDs
16—2N3565 or equivalent transistors
8—220-ohm ½-watt carbon resistors
8—47K-ohm ½-watt carbon resistors

CHAPTER 3

How Digital Integrated Circuits Operate

The electronics industry is into its third generation of major electronic devices. First there were vacuum tubes, then transistors and now integrated circuits (ICs). The amount of circuitry and the numerous complex functions that are built into these tiny packages stagger the imagination and leave you thinking that they border on science-fiction. In fact, it would require a high-powered microscope to see the various sections of the circuitry, once the IC package was broken open. Of course, it is senseless to think about replacing any of the individual (discrete) components and you should think of the IC as a "black box" that will perform some designated function when operated correctly. However, a knowledge of how ICs are fabricated and their basic circuitry will enable you to understand better how to use them.

In this chapter you will learn how ICs are manufactured, basic circuit analysis of TTL integrated circuits used in this book, MOSFET integrated circuits which provide greater packaging density (more circuits per given physical area), IC terminology and some valuable information on handling and connecting ICs.

Project #3 at the end of the chapter involves building the "LOCO" probe which is a test instrument that can be used with the IC breadboard/tester or used to test other digital equipment.

3-1 TECHNIQUES OF FABRICATING A MONOLITHIC IC

The techniques of fabricating a monolithic IC are similar to those used in fabricating an ordinary transistor. These techniques are arranged in a sequence of steps given as:

1. Surface preparation
2. Epitaxial growth
3. Photomasking
4. Diffusion
5. Metallization
6. Packaging

Surface Preparation

Wafers approximately 10 mils thick are cut from a 1 to 2.5-inch silicon crystal ingot as shown in Figure 3-1.

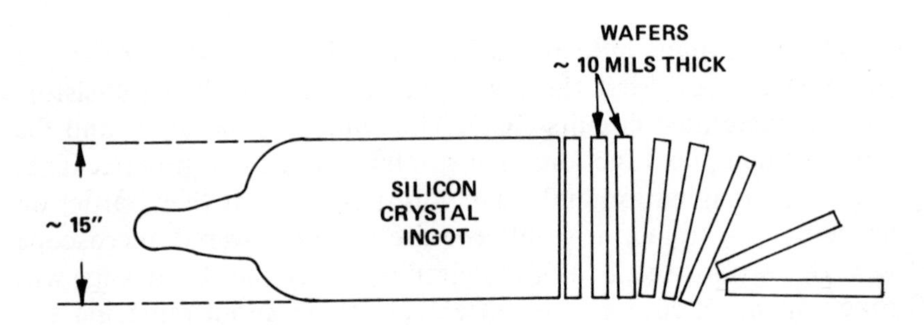

Figure 3-1. IC wafers being cut from silicon ingot *(Courtesy RCA Electronic Components)*

Each wafer is cleaned and polished mirror smooth and becomes the P-type substrate upon which the rest of the IC elements are formed. This P-type substrate acts as the foundation for the IC and also isolates the various elements.

Epitaxial Growth

The polished wafers are now placed into a chamber of approximately 1200°C where N-type dopants are added. An N-type layer from 0.5 to 2 mils thick is epitaxially grown. This epitaxially grown layer has the same crystal structure as the P-type substrate and therefore becomes an extension of the original material. The wafers are now heated in a chamber consisting of an oxygen atmosphere where a layer of silicon oxide (SiO_2) is formed over the N-type layer. This SiO_2 layer acts as an insulating and protective layer against outside impurities as shown in Figure 3-2.

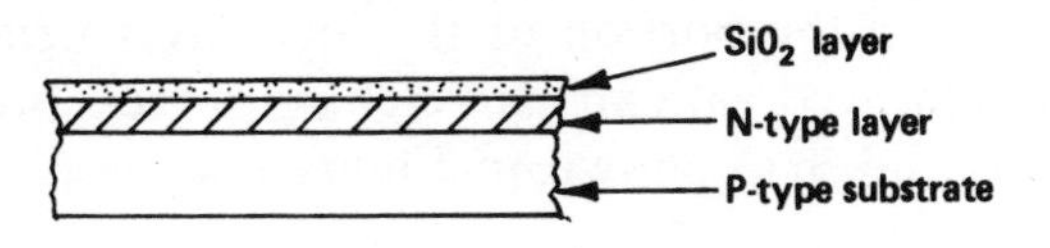

Figure 3-2. Wafer before masking

Photomasking

A portion of the IC circuit defining certain elements is drawn and then through photolithographic techniques is reduced to micro-size prints and placed on a glass plate. This photomask, as shown in Figure 3-3, will determine the inner element patterns fabricated on the IC.

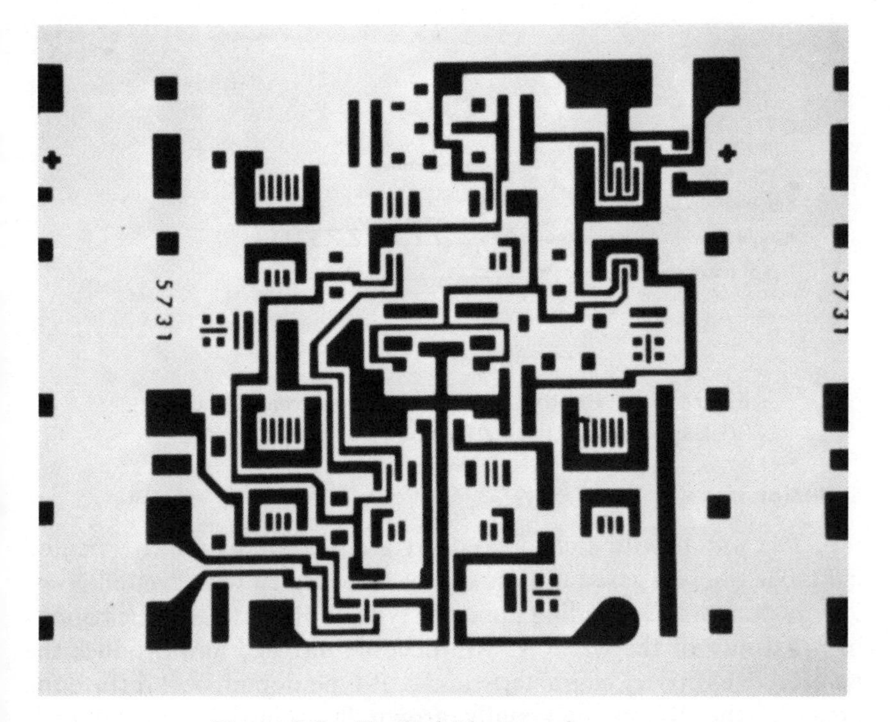

Figure 3-3. Typical Photomask
(Courtesy RCA Electronics Components)

A photoresist chemical is now placed upon the wafer. The photomask is placed above the wafer and exposed to ultraviolet light. The areas of photoresist exposed to the light become hard, whereas the areas of photoresist under the photomask pattern remain soft.

The photoresist and the portion of the SiO_2 layer beneath the soft photoresist are removed with an acid etching process producing windows in the SiO_2 layer as shown in Figure 3-4.

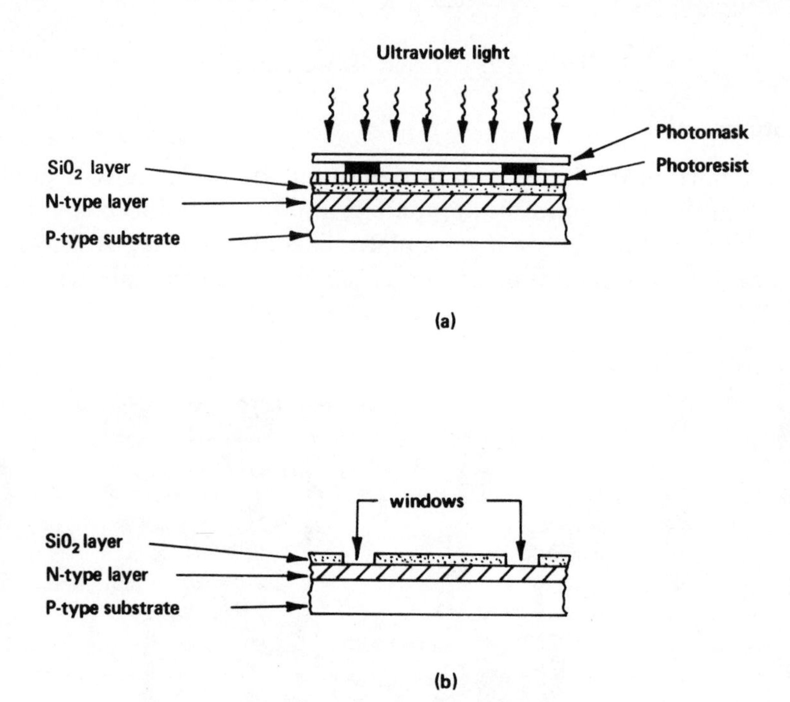

Figure 3-4. Photomasking and Photoresist Processing (a) Exposure (b) Etched windows in wafer

Diffusion

The wafers with etched windows are placed again into a heated chamber where a gaseous atmosphere is saturated with desired N- or P-type dopants. Depending upon the type of dopant used, the regions in proximity of the windows will become diffused and produce the respective electrical characteristics. A P-type dopant will diffuse the areas of the N-type epitaxially grown layer into P-type material which is similar to the substrate. This leaves an isolated area of N-type material which is referred to as an island or boat as shown in Figure 3-5. During this diffusion process an oxygen atmosphere is also passed over the wafer, forming another SiO_2 layer which closes the windows.

These photomasking, photoresist-acid etching, diffusion and oxidation (forming SiO_2 layer) processes are repeated the required

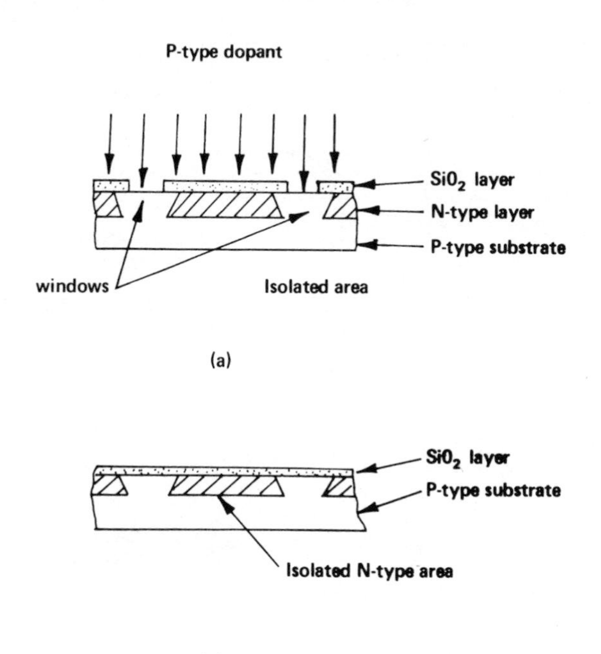

Figure 3-5. Diffusion Process
(a) P-type dopant diffusion
(b) Windows closed with SiO_2 layer

number of times for the desired IC. Figure 3-6 shows the sequences used to fabricate an NPN transistor. A P-type dopant is used to form the base region of the transistor within the isolated N-type collector area as shown in Figure 3-6f. An N-type dopant is then used to form the emitter region within the P-type base region as shown in Figure 3-6h.

Resistors and capacitors are also formed at the same time as transistors and diodes are fabricated. Resistors can be fabricated as shown in Figure 3-7a, where the concentration of dopant, cross-sectional area and length determine the value of resistance. Physical size and practical applications limit the value of resistance up to only a few thousand ohms. Capacitors can be formed by reversed bias diodes or with an N-type layer as one plate, the SiO_2 layer as the dielectric and a metallized area as the other plate as shown in Figure 3-7b. The physical size of the capacitors limits their practicality and they are usually avoided in IC circuitry. The inherent problems associated with inductors have resulted in little or no fabrication of this component into IC manufacturing.

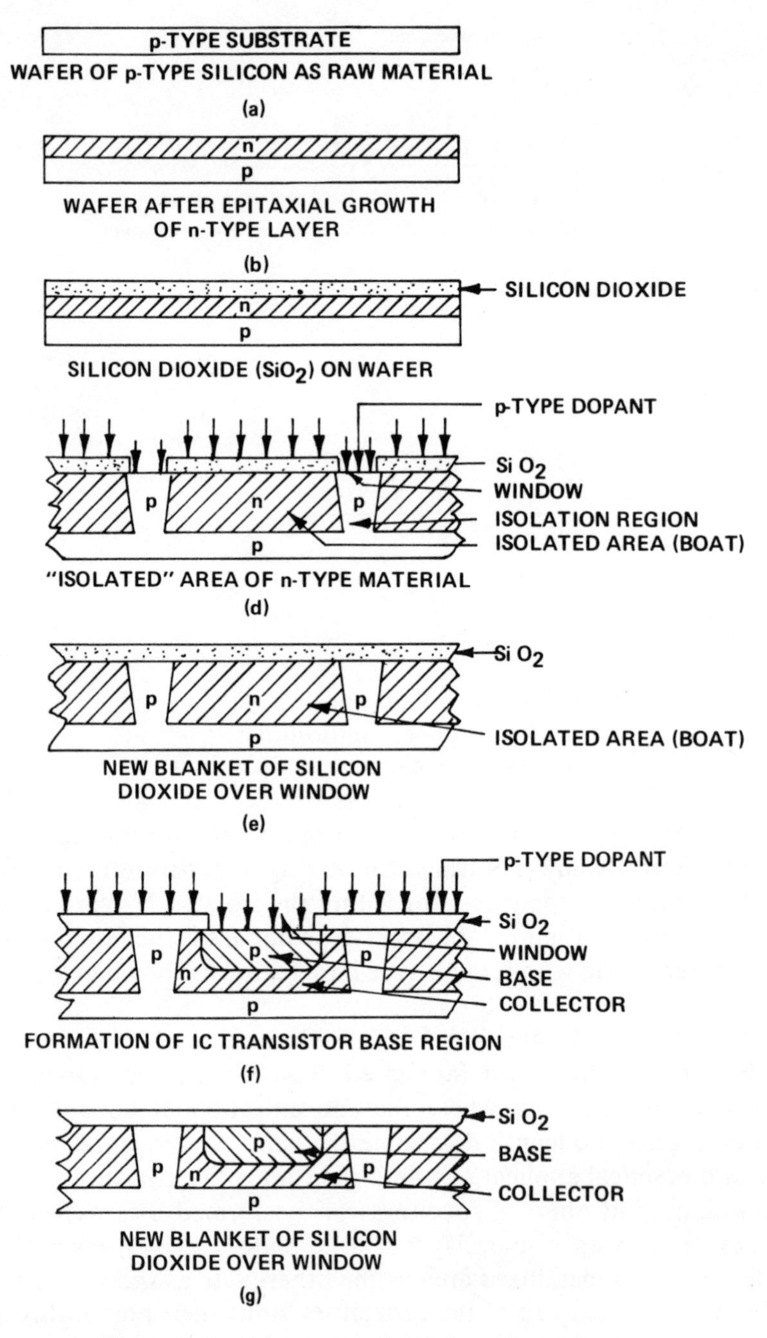

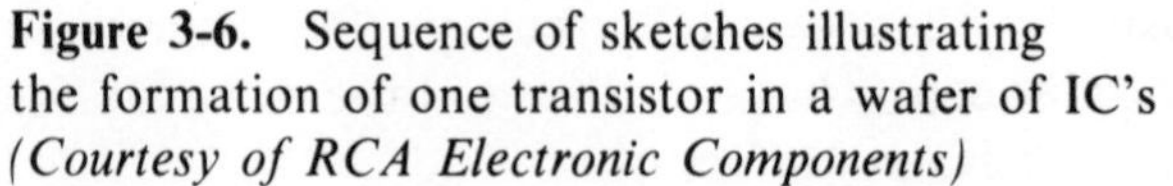

Figure 3-6. Sequence of sketches illustrating the formation of one transistor in a wafer of IC's *(Courtesy of RCA Electronic Components)*

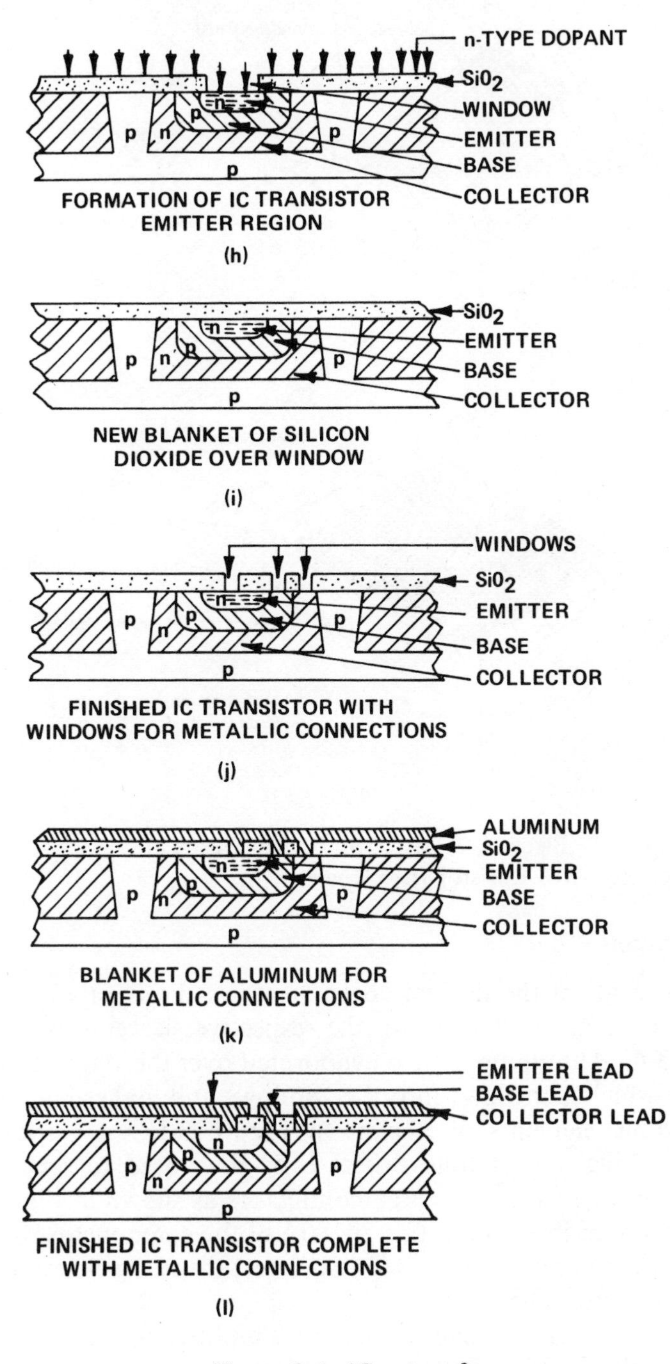

Figure 3-6. (*Continued*)

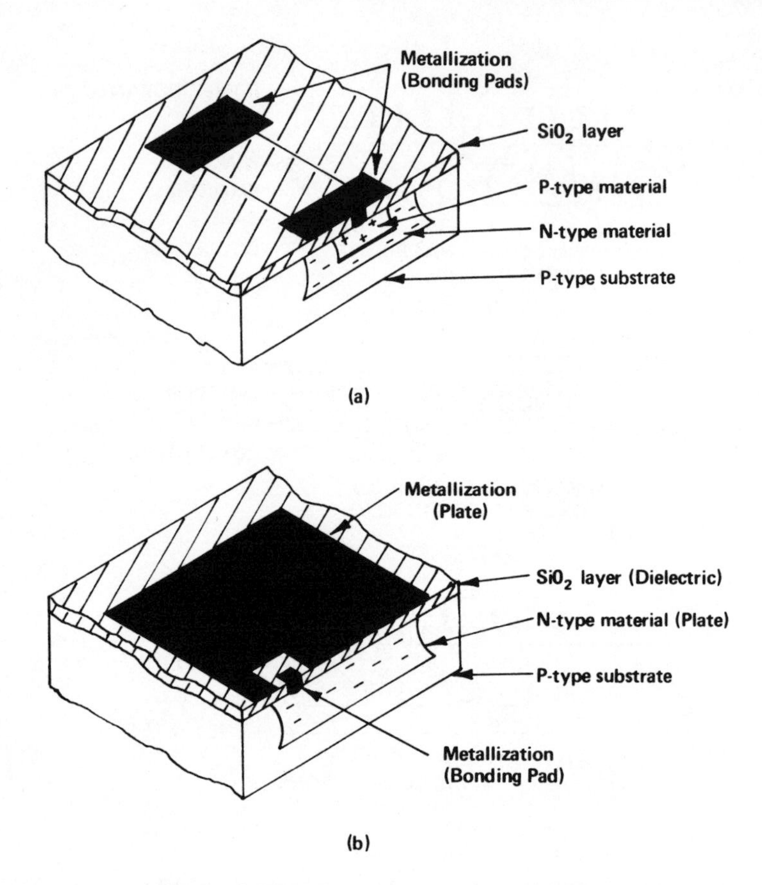

Figure 3-7. IC Component Fabrication
(a) Resistor (b) Capacitor

Metallization

When all of the desired components of the IC are completed, windows are again etched into the respective elements as shown in Figure 3-6j. Aluminum is then evaporated over the wafer, forming a blanket which also flows into the windows and makes contact with the various elements as shown in Figure 3-6k. A variation of photomasking and etching processes is then used to separate the aluminum blanket into isolated conductors as shown in Figure 3-6l. The aluminum interconnection pattern of the components on the IC also includes bonding pads used to connect the IC to external pins as shown in Figure 3-8.

Figure 3-9 shows an example of a simplified monolithic IC chip and the relationship of the fabricated components.

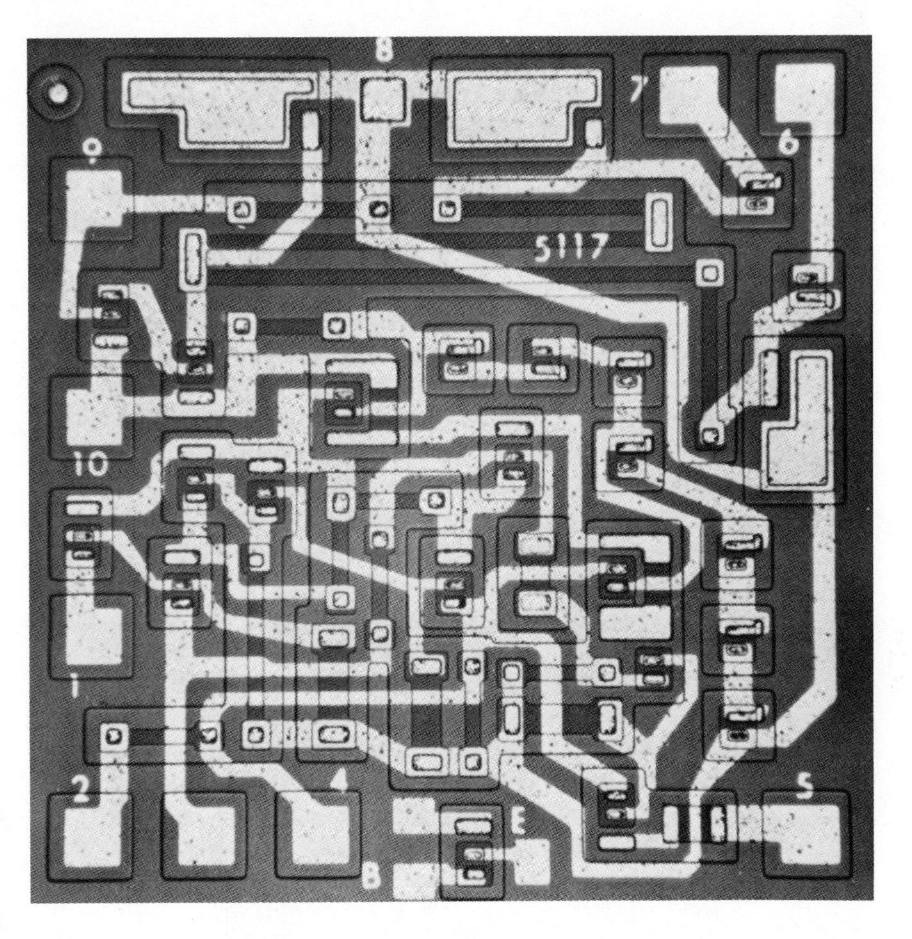

Figure 3-8. IC "chip"
(Courtesy RCA Electronic Components)

Packaging

Each complete IC ranges in size from 15 mils to 100 mils square and is referred to as a "chip." From 250 to over 1,000 individual chips can be fabricated simultaneously on a single wafer depending upon the IC's complexity as shown on the finished wafer in Figure 3-10.

The individual chips are initially tested by automated test equipment while still in the wafer form. Those chips which do not pass the required tests are marked and discarded later. A cutting operation using a diamond-tipped scriber separates each chip from the wafer. Those chips which tested good are then mounted in a package where

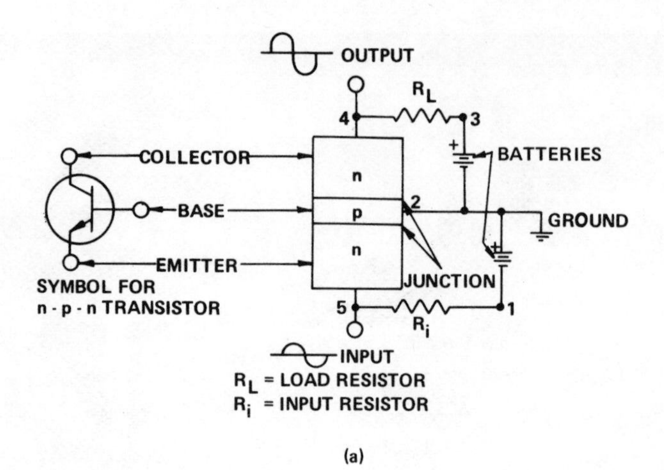

(a)

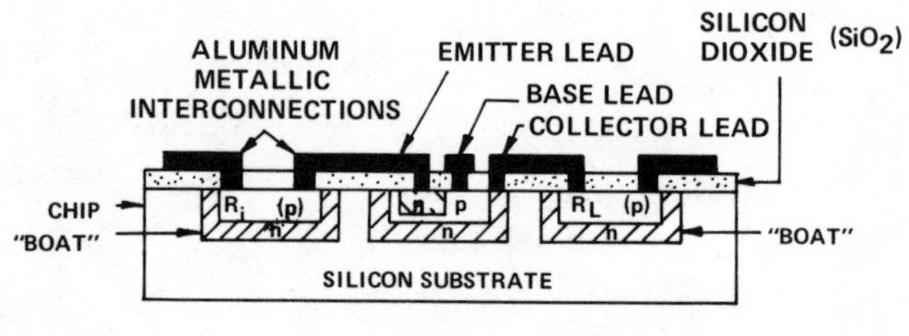

(b)

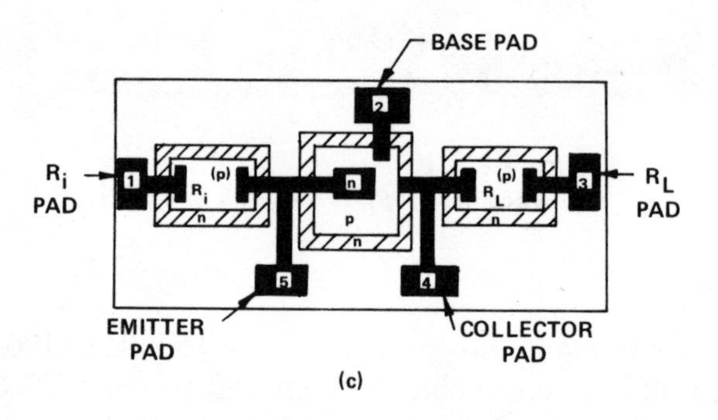

(c)

Figure 3-9. Simplified Monolithic IC Chip
(a) Schematic Diagram (b) Cross-section view
(c) Top View *(Courtesy RCA Electronic Components)*

gold or aluminum wire leads with approximately 1.5-mil diameters are connected between the bonding pads of the IC to the appropriate terminal posts of the package as shown in Figure 3-11.

The cap is placed on the package and is sealed a variety of ways,

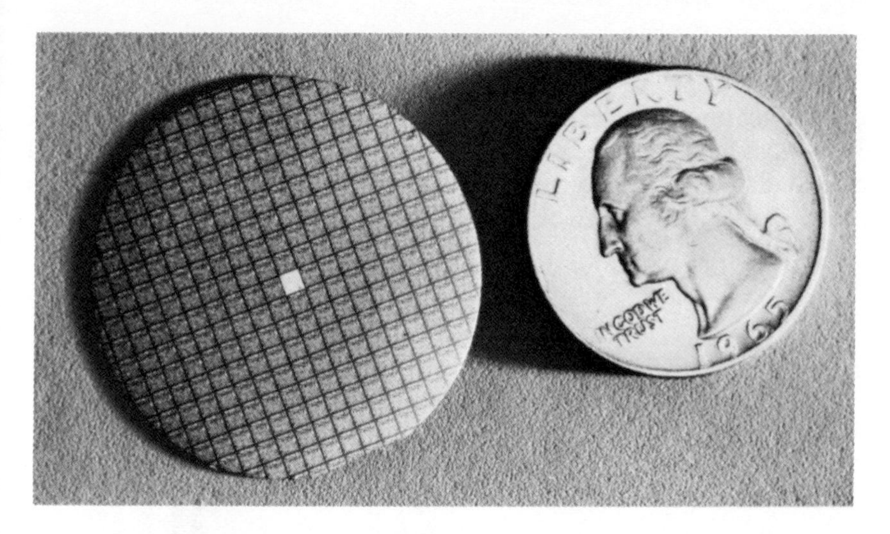

Figure 3-10. A Wafer of Finished IC Chips
(Courtesy RCA Electronic Components)

Figure 3-11. IC Chip mounted in TO-5 package assembly
(Courtesy RCA Electronic Components)

depending upon the design and type of material used for the package. The IC is now tested for any "leaks" in the package and a final electrical test results in a completed IC. Integrated circuit packages are available in a variety of styles and shapes as shown in Figure 3-12.

Figure 3-12. IC Packages: ("TO-5 can," flat-pack and dual-in-line types *(Courtesy RCA Electronic Components)*

3-2 TRANSISTOR-TRANSISTOR LOGIC (TTL)

The 7400-series integrated circuits use transistor-transistor logic (TTL or T^2L) which refers to the type of electronic components used in constructing logic gates. Other types of logic will be explained in Section 3-4 under IC terminology.

The 7400 TTL logic is currently the most popular, largely

because it has the second-highest operating speed of all logic types, has low power dissipation, high fan-out capabilities and excellent noise margin.

Although there are a number of IC manufacturers, most of them generally adhere to a standard form of identification. The 74 means it is suitable for the industrial temperature range between 0° and +70°C, while the remaining numbers designate the type of logic gate or device that it is. The 5400 series IC is suitable for the military temperature range between −55° and +125°C. A letter between these two sets of numbers may designate whether it is (H) high speed, (L) low speed, or (S) "Schottky diodes" which have been incorporated into the circuitry to increase the operating speed. No letter between these numbers means that it is standard. Letters preceeding the numbers usually identify the type of component that it is, for instance, SN simply stands for integrated circuit or DM stands for digital monolithic. A letter following the numbers usually refers to the type of package that the IC is mounted in or gives other specifications. The following example may help clarify IC identification.

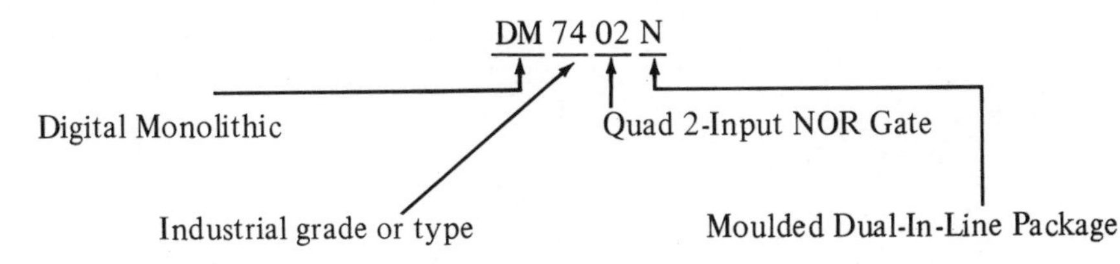

The basic TTL gate is the NAND gate shown in Figure 3-13. There are four of these separate identical circuits in a 7400 Quad 2-Input NAND gate package. Nearly all other TTL gates have a similar configuration.

Transistor Q1 has multiple emitter inputs and functions as an AND gate that is directly coupled to Q2 which serves as the inverter producing the NAND gate. Components R3, Q3, D1, and D4 act as a buffer and switching arrangement which allows the output to be either high (approximately +3.6V) or low (approximately ground). When Q3 conducts, the output is high and when Q4 conducts, the output is low. Diode D1 doesn't permit Q3 and Q4 to conduct at the same time. This output circuit is often referred to as a "totem-pole" output and enables the gate to drive 10 other gates or devices satisfactorily.

In most cases diodes (not shown) are connected from the input emitters with their anodes to ground to serve as input protection against high-voltage transients. The transistors and resistors are so designed that the condition of Q1 controls the condition of Q2. In turn, Q2 controls the condition of Q3 and Q4.

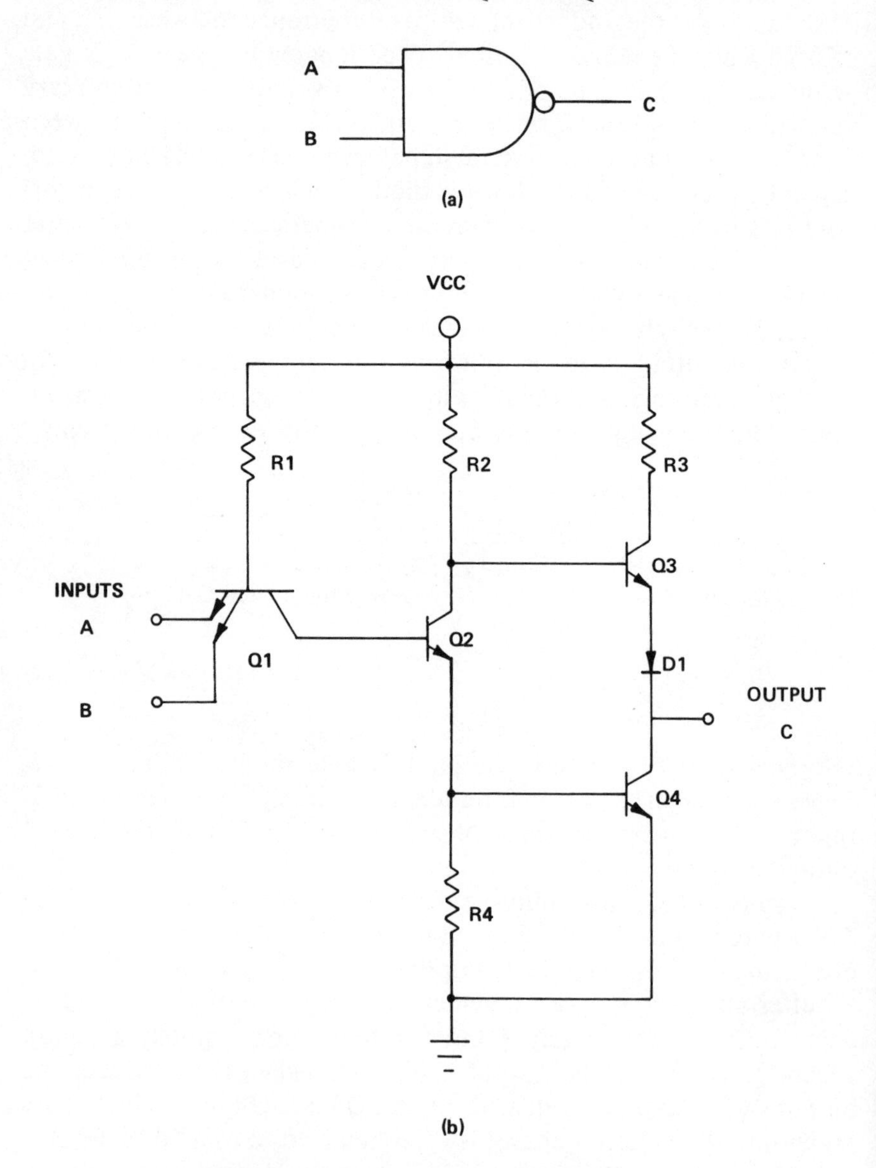

Figure 3-13. IC TTL NAND Gate
(a) Logic Symbol (b) Schematic Drawing

When All Inputs Are High. (Figure 3-14a)

When all inputs are high, Q1 is cut off and the current through R1 (conventional flow) flows into the base of Q2, turning it on. The

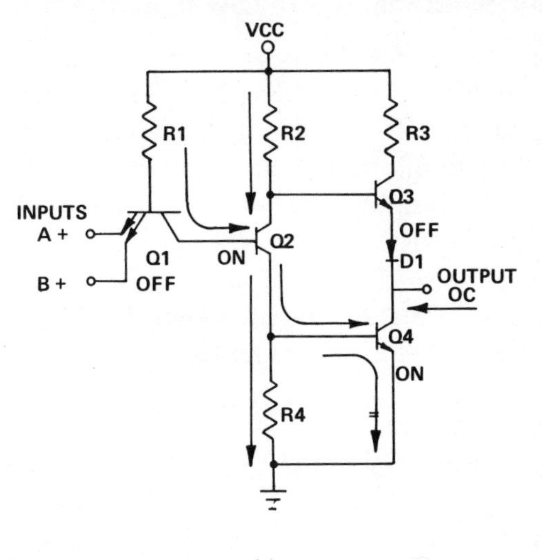

(a)

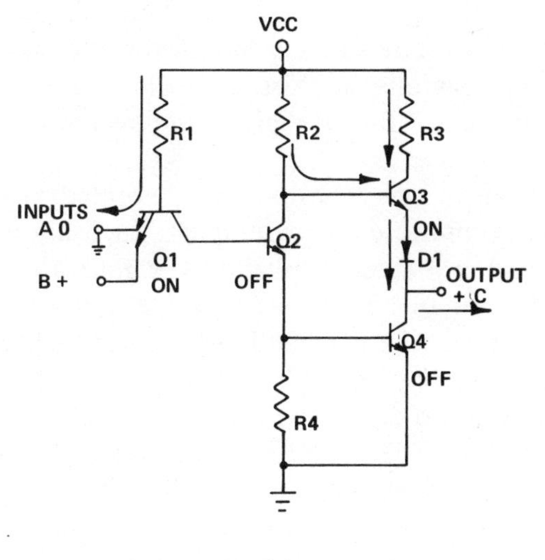

(b)

Figure 3-14. TTL NAND Gate Operation
(a) When all inputs are high
(b) When any one or all inputs are low

lower voltage on the collector of Q2 cuts off Q3 and the full current of R2 flows down through Q2. The voltage drop developed across R4 turns on Q4 which "sinks" or pulls the output toward ground.

When Any One or All Inputs Are Low (Figure 3-14b)

When any one or all inputs are low, Q1 turns on allowing the current through R1 to flow to ground. The lower voltage on the collector of Q1 turns off Q2. The higher voltage on the collector of Q2 turns on Q3 and allows the current through R2 and R3 to flow through Q3 to the output. The output is "pulled up" toward Vcc through D1, Q3 and R3.

3-3 METAL OXIDE SEMICONDUCTOR FIELD-EFFECT TRANSISTOR (MOSFET)

The metal oxide semiconductor field-effect transistor (MOSFET) operates very much like the vacuum tube as shown in Figure 3-15.

Two P-type regions are diffused side by side into an N-type substrate and become the source (tube's cathode) and the drain (tube's plate). An SiO_2 layer is formed over the surface and serves as insulation between the two regions. Two areas of metallization are made to penetrate through windows in the Si02 layer and contact the drain and source P-type regions. A third area of metallization is formed above the gap in the two P-type regions and becomes the gate (tube's grid).

The operation of the MOSFET is illustrated in Figure 3-16.

The source is connected to ground and the drain has a -15 volts applied to it. With a -1 volt applied to the gate, a few holes are attracted toward the gap between the source and drain and form a small P-type channel, where a small hole current is produced. When the gate voltage becomes more negative, more holes are attracted toward the gap and the size of the P-type channel increases, creating a larger current flow. The MOSFET performs much the same function as a transistor except that it is controlled by voltage similar to a vacuum tube rather than by current like a bipolar transistor. Since no channel exists without proper gate voltage, this MOSFET is called the enhancement-mode type. Because of the SiO2 insulating layer and the resulting capacitive action the gate has a high input impedance (approximately 10^{12} ohms).

PLATE

GRID

CATHODE

(a)

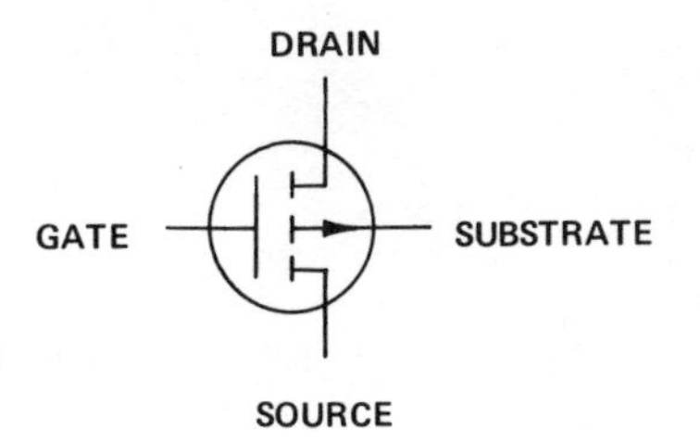

(b)

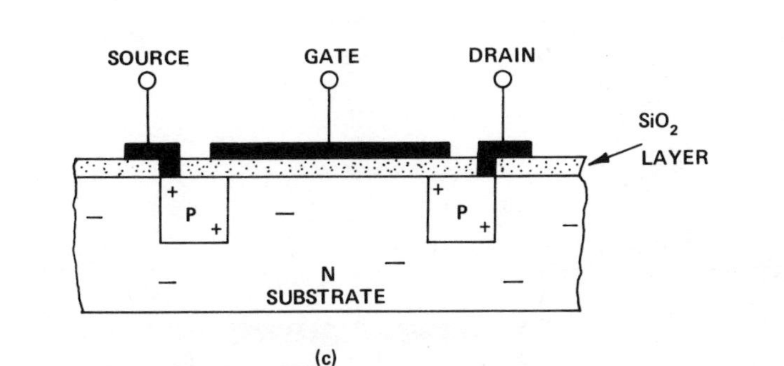

(c)

Figure 3-15. P-Channel MOSFET
(a) Vacuum Tube Schematic Diagram
(b) P-Channel MOSFET Schematic Diagram
(c) Cross-section View of Chip

There is another type of MOSFET called "depletion-type," where the channel is permanently diffused into place during fabrication. This depletion-type MOSFET is not as well suited to IC's as the enhancement-type MOSFET.

Some manufacturers simplify the MOSFET schematic diagram as shown in Figure 3-17a.

A MOSFET that is specially controlled during the fabrication process and/or uses a fixed gate voltage (V_{GG}) can serve as a resistor as compared in Figure 3-17b and c. An alternate method of representing a MOSFET used as a resistor is shown in Figure 3-17d.

CMOS

Another area of MOS technology uses P-channel and N-channel MOSFET's fabricated on the same chip in a complementary

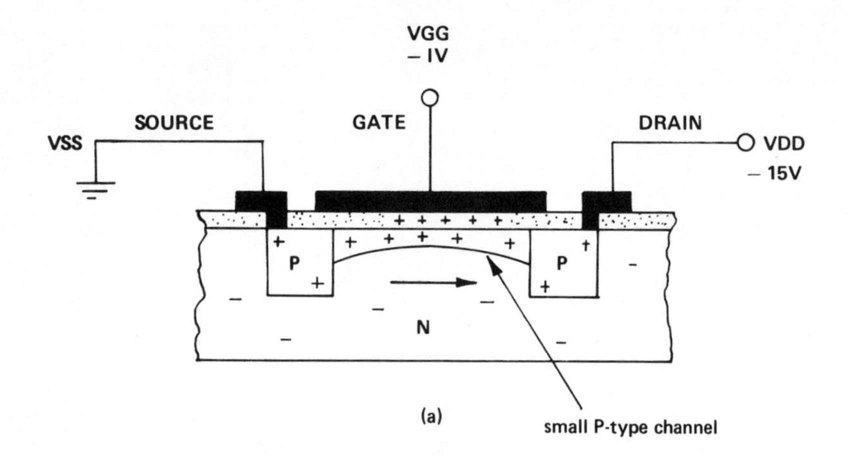

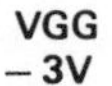

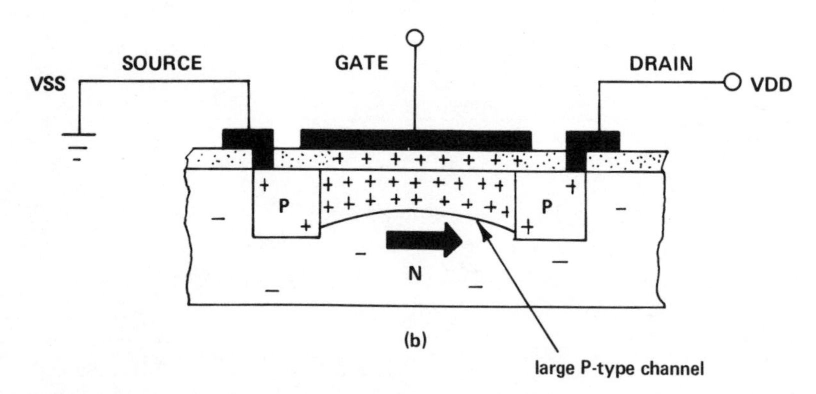

Figure 3-16. MOSFET Operation (a) Small Gate bias produces small channel (b) Large Gate bias produces large channel

switching arrangement and is called CMOS or like the popular RCA COS/MOS (Complementary-Symmetry MOS) CD 4000 series. Some CMOS ICs are pin-for-pin compatible with the TTL 7400 series ICs.

Advantages of MOSFET ICs

MOSFETs require less space than bipolar transistors which permits greater packing density and complexity on a single chip. An example of this is the microprocessor chips which are finding their way into scientific pocket calculators. MOSFET ICs are also used quite extensively in semiconductor memories which demand storage capacity for many thousands of bits.

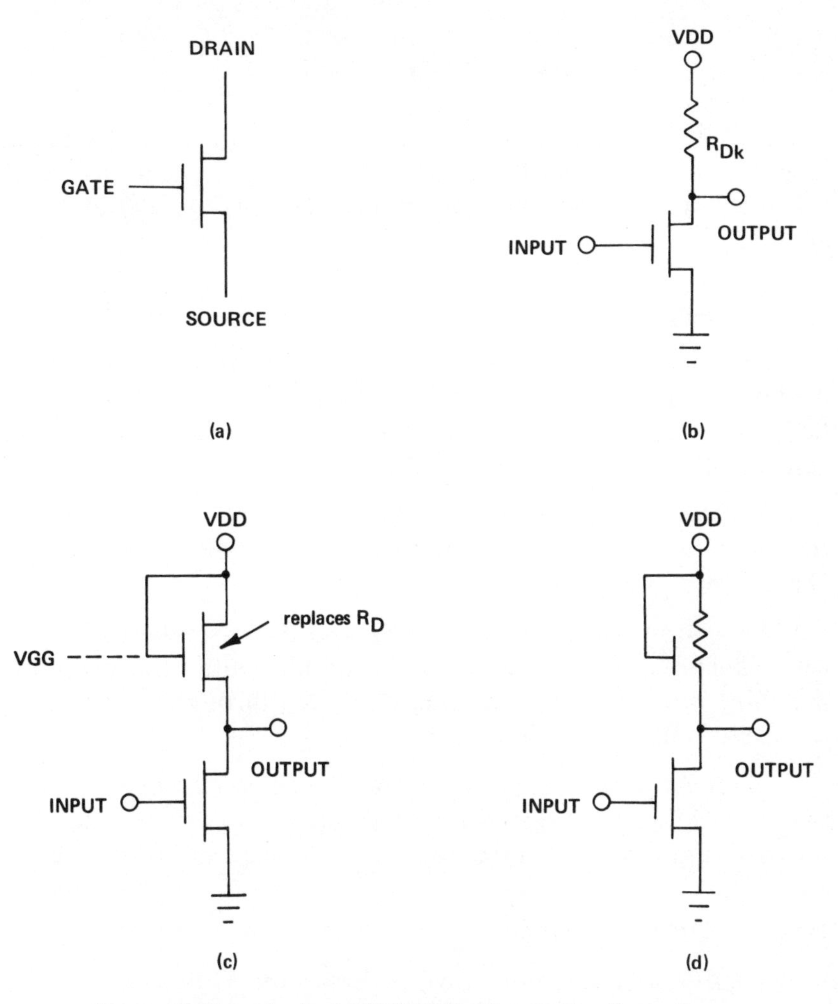

Figure 3-17. Typical MOSFET Circuit Configurations (a) Simplified Schematic Diagram Symbol (b) Common Source Amplifier (c) MOSFET used for R_D (d) Alternate Schematic Diagram

The circuits used in MOSFET ICs are simpler to manufacture than TTL ICs since they virtually do not require resistors, diodes, bipolar transistors or other components.

Only one diffusion process is required of MOSFET ICs as opposed to three diffusion processes with TTL ICs. The single-diffusion process is required of MOSFET ICs as opposed to three diffusion processes with TTL ICs. The single-diffusion process is less costly and the yield of good chips is higher because the repeated diffusion process with TTL tends to introduce defects.

3-4 IC TERMINOLOGY

This section lists some of the more commonly used terms associated with the use of ICs and their fabrication. The terms are not necessarily in alphabetical order, but are grouped in relation to one another.

Types of Logic

Diode-Transistor Logic (DTL): an early type of logic using diodes, transistors and resistors. The diodes produce the AND function, while the transistor serves as the inverter to produce the NAND gate which is the basic gate for DTL.

Resistor-Transistor Logic (RTL or TRL): uses only resistors and transistors where in a sense the resistors have replaced the diodes of DTL to produce the basic NOR gate.

Transistor-Transistor Logic (TTL or T^2L): is similar to DTL except the input diodes are replaced by multiple-emitter transistors and matching problems are appreciably reduced through isolation with the use of output switching transistors.

Emitter-Coupled Logic (ECL) or Current-Moda Logic (CML): this type of logic is designed so the transistors do not operate in cut off or saturation which increases the switching time of the gates.

Types of Circuit Integration

Small-Scale Integration (SSI): refers to the IC Process when less than 10 logic elements are fabricated on a single chip.

Medium-Scale Integration (MSI): refers to the IC process when less than 100 logic elements are fabricated on a single chip.

Large-Scale Integration (LSI): refers to the IC process when more than 100 logic elements are fabricated on a single chip.

Types of ICs

Digital IC: an IC used for switching or dealing with binary information.

Linear IC: an IC used for amplifying or regulating hence, sometimes called an analog IC.

Hybrid IC: an IC that is not monolithic, but made up of two or more different technologies where the separate parts are mounted on an insulating substrate and connected together by wires or a metallization pattern.

IC Fabrication Terms

Wafer: a thin piece of semiconductor crystal (silicon) approximately 3 inches in diameter, upon which many IC's are formed during the fabrication process.

Die or Chip: a small part of a wafer of silicon upon which an IC is fabricated.

Substrate: an insulating crystalline material support upon which an IC is fabricated or mounted.

Epitaxial Growth: the process of chemically extending and modifying a crystalline substrate by the addition of dopants in an epitaxial furnace where the material grows in a precise and controllable manner.

Dopants: impurities that are introduced under highly controlled conditions to a wafer of silicon to change its electrical characteristics.

Diffusion: the process of introducing small quantities of dopants in a highly controlled manner into a crystalline material to modify its electrical characteristics.

Isolation Diffusion: the process of diffusion to separate the discrete components within a crystalline structure.

Land: a specific area or "island" of special material isolated within another material by means of isolated diffusion.

Photoresist: a liquid material which hardens into an acid-resistant solid when exposed to ultraviolet light.

Photomask: a transparent plate, containing opaque circuit configurations, which is used with photolithographic techniques in fabricating IC's.

Etching: the removal of surface material, unprotected by hardened photoresist, from a wafer by chemical means, usually acid.

Window: a small area cut into or through a layer of material to produce other crystalline structures by diffusion or to make contact by metallization.

Metallization: the process of evaporating aluminum over a wafer so that an interconnection pattern can be made for connecting the various components of an IC.

Bonding: the process of joining together two types of material, such as the wires from the pin connections to the bonding pads of a chip.

Mil: a unit of length equal to one-thousandth of an inch.

Micron: a unit of length equal to one-millionth of a meter. One micron equals approximately .04 mils. The SiO_2 layer on a chip is about 1 micron thick.

Monolithic: a material with different electrical characteristics, but with the same crystalline structure throughout.

Planar: refers to the technique in which the various semiconductor junctions are brought to a common surface of a chip.

MOS: stands for metal oxide semiconductor.

FET: stands for field effect transistor where this semiconductor device is controlled by voltage rather than current.

MOSFET: stands for metal oxide semiconductor field effect transistor.

IGFET: stands for insulated gate field effect transistor and is synonymous with MOSFET.

JFET: stands for junction field effect transistor.

CMOS: stands for complementary metal oxide semiconductor.

COS/MOS: stands for complementary-symmetry metal oxide semiconductor.

Density Packing: refers to the amount of elements that can be placed within a specific area. LSI has a high-density packing.

IC Packaging

TO-CAN: stands for transistor-outline metal-can package. The TO-CAN is about ⅜ inches in diameter and resembles a standard transistor can. It may have 8, 10 or 12 wire pins projecting from its underside. IC sockets are designed for these packages.

Flat-pack: a thin rectangular IC package with the pins (or leads) coming straight out of the sides. These packages are nearly always wired hard into a circuit.

DIP: stands for dual-in-line package. This rectangular package is thicker than the flatpack, with its pins coming out only on two sides of the package parallel to each other and bent downward. The pins are designed with shoulders which hold the package away from the printed circuit board. This feature allows some heat dissipation and permits wires to be run beneath the package. Some of the more common types of these packages contain 14, 16, 24 and 28 pins spaced 0.1 inch between centers. IC sockets are available for these packages.

IC Operating Characteristics and Features

Unipolar: a transistor which has charge carriers of a single polarity resulting in current flow in one direction. The MOSFET is a unipolar transistor.

Bipolar: a transistor which has majority and minority carriers resulting in current flow in two directions. The junction transistor is a bipolar transistor.

Current Source: the conventional current flow from a positive potential or source. An example of current sourcing is shown coming out of the output of the circuit in Figure 3-14b.

Current Sink: the component that allows conventional current to pass to ground. An example of current sinking is shown going into the output and through Q4 to ground of Figure 3-14a.

Current Hogging: may occur when one of several driven circuits "hogs" too much current, not allowing the other circuits to function properly.

Fan In: the number of inputs that can safely be connected to a logic circuit.

Fan Out: the number of parallel loads that can safely be driven from a logic circuit. The fan out of most 7400 series IC's is 10 per circuit.

Power Dissipation: the amount of power that an IC can handle and safely dissipate the resulting heat. Typically, the 7400 series IC may dissipate about 10 milli-watts per gate.

Military-grade (5400 Series) IC: an IC that is guaranteed to operate over the temperature range from −55 to +125° centigrade.

Industrial Grade IC (7400 Series) IC: an IC that is guaranteed to operate over the temperature range from 0° to 70° centigrade.

Propagation Delay: the time that it takes a digital signal to travel through a logic device or logic system, usually measured in nanoseconds.

Threshold Voltages: the minimum and maximum voltage levels which determine the logical state of a digital circuit.

Noise Margin: the transient voltage level between the normal operating logic levels and the threshold voltage in which an IC will not operate falsely.

Noise Immunity: same as noise margin.

V_{CC}: the supply voltage connected to the IC for proper operation. The V_{CC} for the 7400 series IC is set at +5 volts.

V_{SS}: the supply voltage connected to the source of an FET.

V_{GG}: the supply voltage connected to the gate of an FET.

Pull-up: refers to a resistor or transistor which is connected between the positive supply voltage and the output of a logic circuit.

Pull-down: refers to a resistor or transistor which is connected between ground and the output of a logic circuit.

Open Collector Output: an IC in which the final pull-up resistor in the output transistor is omitted during fabrication and an external resistor must be provided before the circuit is complete.

Wired-OR Circuit: usually open collector output ICs whose outputs are wired together to form the OR function. The circuit requires at least one pull-up resistor and the output will be logic 0 if one of the input circuits is in a logic 0 state.

Data Bus: a common electrical path over which digital information is transferred from any one of several circuits or sections to another. Only one transfer of information is possible at a time and all other sources that are tied to the bus must be disabled during the transfer.

Tri-state Logic® : trademarked by National Semiconductor and also called; three-state logic, three-state TTL and three-state outputs. A special IC which is designed to have three possible output states, i.e.; a logic 0 state, a logic 1 state and a state in which the output is disconnected from the rest of the circuit. Tri-state logic® is particularly useful with buses.

A portion of a typical IC information sheet which uses some of the terminology given is shown in Figure 3-18.

3-5 WIRING, SOLDERING AND HANDLING ICS

Integrated circuits are encapsulated in fairly rugged packages: however, a certain amount of care is required to protect them and insure their reliability. Semiconductor devices are very susceptible to damage from excessive voltage current and heat. The absolute maximum voltage ratings for ICs given by the manufacturer should never be exceeded, therefore, it is very necessary to have a power supply with nearly perfect regulation.

Storing ICs

The pins or leads on ICs are very delicate and can not be bent very often before breaking. Naturally, extreme care should be taken when handling ICs, but when any electronic components with leads are stored in drawers or bins, inevitably the leads become bent, intertwined or even broken. There are individual plastic containers made for ICs, but this could be a little costly if you intend to keep many ICs on hand. One inexpensive method of storing ICs is to use styrofoam about ⅜ to ½ inch thick. Needless to say, the ICs are simply pushed (pins down) into the styrofoam. The styrofoam may be cut a little larger than the IC it is to hold or a large panel (say 6 by 8 inches) may be used to hold all of the ICs in a planned sequence according to their numbers. This latter technique of course makes it easier to locate a specific IC when one is needed for replacement or building projects. In the event some pins become bent, they can very easily be restored by carefully using a pair of long-nose pliers.

Soldering ICs

The principal consideration when soldering ICs is to remember not to use too much heat for too long a period of time. The soldering iron used on ICs should be rated at 50 watts or under. Using heat-

Series 54/74

DM5400/DM7400(SN5400/SN7400) quadruple 2-input NAND gate
DM5410/DM7410(SN5410/SN7410) triple 3-input NAND gate
DM5420/DM7420(SN5420/SN7420) dual 4-input NAND gate

general description

Employing TTL (Transistor-Transistor-Logic) to achieve high speed at moderate power dissipation, these gates provide the basic functions used in the implementation of digital integrated circuit systems. Characteristics of the circuits include high noise immunity, low output impedance, good capacitive drive capability, and minimal variation in switching times with temperature. The gates are compatible with and interchangeable with Series 54/74 equivalent.

features

- Typical Noise Immunity 1V
- Guaranteed Noise Immunity 400 mV
- Fan Out 10
- Average Propagation Delay 13 ns
- Average Power Dissipation 10 mW per gate

schematic and connection diagrams

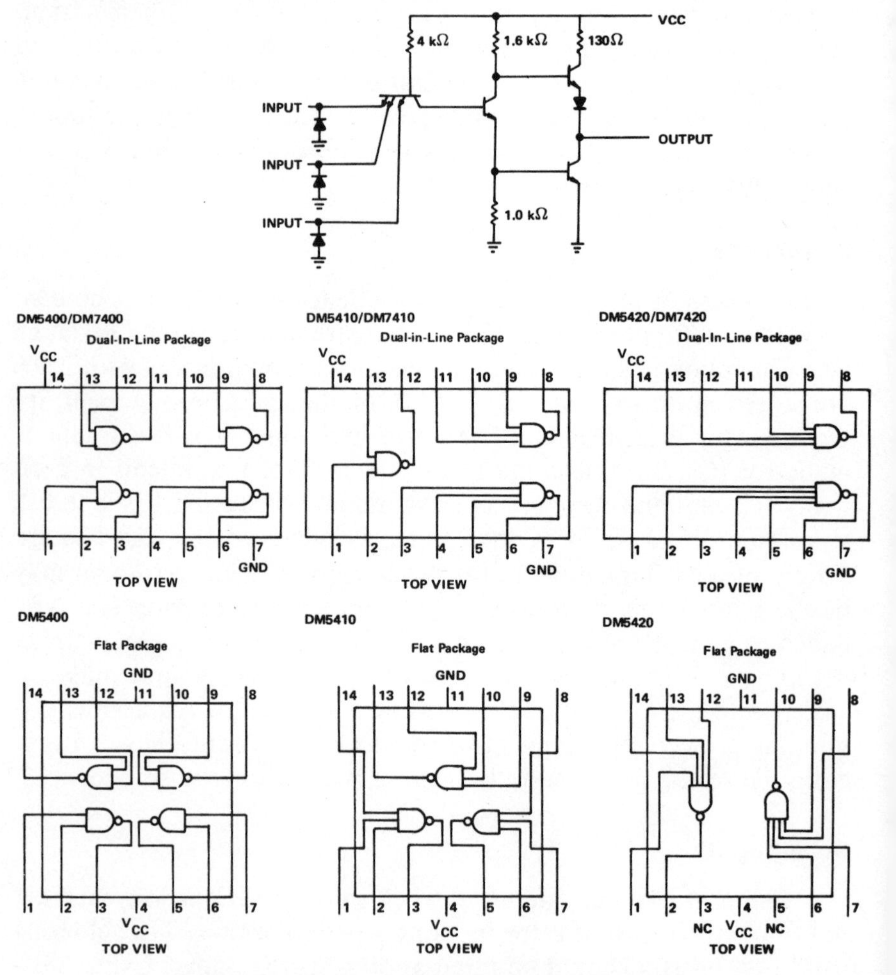

Figure 3-18. Typical IC Information Sheet
(Courtesy National Semiconductor Corporation)

sink techniques as with transistors is inefficient because the IC has many leads which are spaced close together. Figure 3-19 shows some methods of soldering ICs to printed circuit (PC) boards.

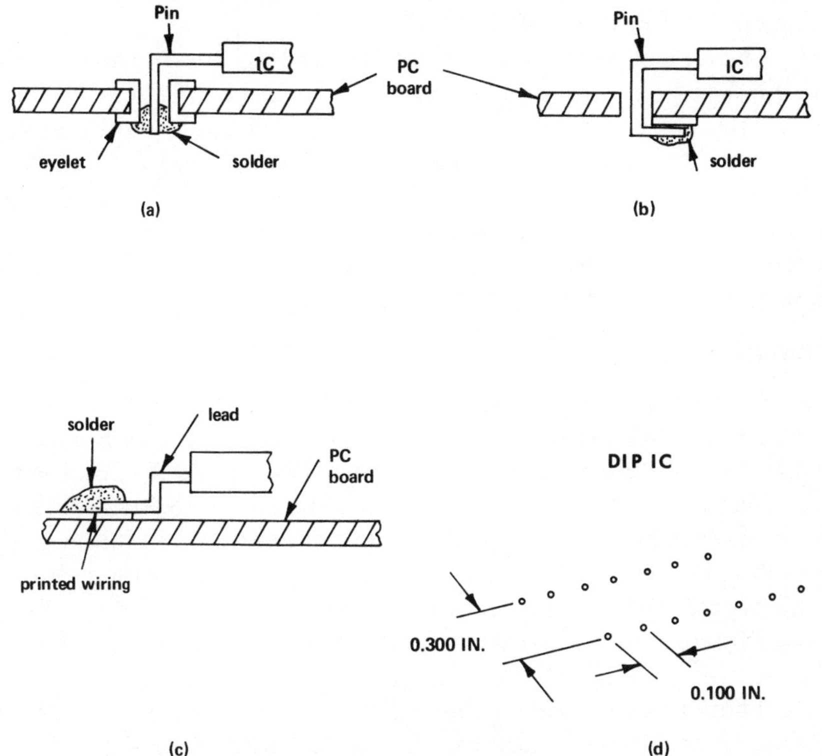

Figure 3-19. IC Soldering Techniques
(a) Straight-through Method
(b) Bending Method (c) Surface-connection Method (d) DIP Mounting Method

Factory-assembled boards usually use the straight-through method as shown in Figure 3-19a, and the leads are connected simultaneously by dip soldering or wave soldering. This method is difficult during repair because the PC board must be turned over and the IC held in place while soldering each pin. A short bend in each pin, as shown in Figure 3-19b, will sufficiently hold the IC in place and make soldering easier.

Some PC boards may have the ICs mounted with the surface-connection method as shown in Figure 3-19c. In this case it may be required to bend the pins. Extreme caution must be used in bending pins so as not to break the seal where the pins enter the IC package. The safest way to accomplish bending is to use two pairs of long-nose pliers, one pair to support the pin while the other pair performs the actual bending. The radius of the bend should not be less than the diameter or thickness of the pin.

The typical pattern dimensions for a DIP IC are shown in Figure 3-19d. The holes are approximately .04 inches in diameter.

After an IC has been soldered to a PC board, it is important that the connections should be cleaned with alcohol and a small hair-bristle brush (or toothbrush). This will prevent any paths of foreign material between the connections from shorting out, causing faulty circuit operation and possibly destroying the IC. A soldering aid is also useful for removing stubborn residue.

Unsoldering ICs

Removing ICs from a PC board is a difficult and time-consuming job. There are, however, some tools and techniques that make the job easier. There are commercially made IC desoldering tips such as shown in Figure 3-20a. This tool connected to a soldering iron makes contact with all of the pin connections simultaneously enabling the IC to be pulled out of the board with a rocking motion. This type of desoldering tool should not be used to solder a new IC into a board since the excessive time the iron is applied could destroy the IC.

There is also a special tool called a solder gobbler in which the solder on each pin connection is heated and when it becomes molten is drawn away by a vacuum bulb. A simple solder gobbler is shown in Figure 3-20b, which can be used with a standard soldering iron. First the soldering iron is placed on the pin connection. When the solder becomes molten, the iron is removed and with its bulb squeezed the gobbler nozzle is placed on the connection. The bulb is then released, which draws the solder into the nozzle. The gobbler is removed from the connection and the solder expelled making it ready for the next pin connection. When all of the connections are finished, the IC can be removed, usually with some additional application of the soldering iron.

Another method of unsoldering ICs is with the touch-and-wipe technique. A hair-bristle soldering bursh is used in this method. The pin connection is first heated with a soldering iron and the brush is used to wipe the solder away. After a few touch-and-wipe operations, you can proceed to the next pin connection. After all pin connections are reasonably clean, additional heat is usually required to remove the IC.

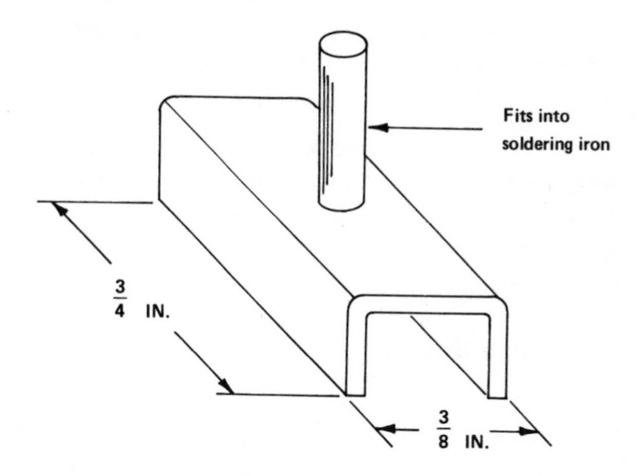

(a)

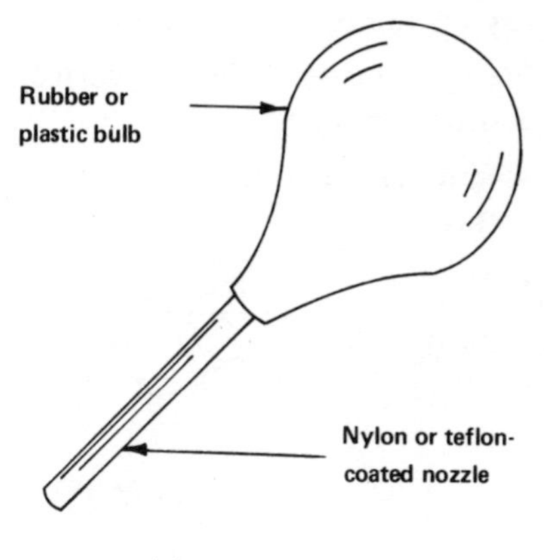

(b)

Figure 3-20. Desoldering Tools (a) Desoldering Tip (b) Simple Solder Gobbler

Installing an IC into a Socket

The following procedure is for installing a DIP IC in a socket, but with some modifications will work well for any IC.

1. Make sure the IC is facing the right direction corresponding to the circuit.
2. Line up all of the pins on one side of the IC with the corresponding holes on one side of the socket.
3. Press these pins slightly into the holes.
4. Line up the pins on the other side of the IC with the corresponding holes in the socket. It may be necessary to bend some pins gently toward the holes with a pointed object.
5. After all of the pins are lined up, gently press the IC into the socket with a rocking motion.

Removing an IC from a Socket

Commercially-made IC pullers are available at parts supply houses or you can make one from a pair of long tweezers by bending the tips ⅛ inch inward 90 degrees (facing each other). A simpler method can be followed:

1. Place the blade of a small screwdriver at one end of the IC between the body of the IC and socket.
2. Pry up the IC a small amount.
3. Place the blade of the screwdriver at the other end of the IC and pry up slightly.
4. Alternately continue this action until the IC is free of the socket.

If you try to pry the IC out of the socket with a single action, the pins on the other end of the IC will be damaged as it comes out of the socket.

SUMMARY

The manufacturing of ICs is a highly precise and complicated process and the functions that these micro-circuits perform are phenomenal. The quality of the IC is maintained from the initial preparation of the substrate throughout the epitaxial growth, photomasking, diffusion, metallization and packaging processes.

Micro-electronics has opened up a new era for Man. The techniques of SSI, MSI and LSI have made ICs a necessity in everyday life. The bipolar TTL ICs, currently so widespread, will eventually be replaced by unipolar MOSFET ICs. Where electronics technology will lead in the future still remains a mystery, but it will be a rewarding challenge as it has been in the past.

It could be said that the person with a knowledge of electronics is a person of the times. By no means, does this relieve the technician from the responsibility of keeping abreast of the sweeping changes in electronics. The electronics technician must learn new technologies, new terminology and new skills in handling these improved devices.

3-6 PROJECT #3: BUILDING THE "LOCO" (LOGIC CONDITION) PROBE

The LOCO probe is a simple but effective test instrument that can be used to determine the LOgic COndition (hence "LOCO") of a digital circuit. The circuitry uses a high input impedance Darlington pair amplifier, an LED output indicator and a one-shot multivibrator IC. (Multivibrators are covered in Chapter 4). The schematic diagram and component layout are shown in Figure 3-21.

Theory of Operation

In normal operating condition, when the input signal at the probe is 0, the Darlington pair is cut off and the LED is extinguished. When a 1 is present at the probe the Darlington pair turns on, providing a path for current to flow through the LED, which in turn glows giving an output indication of the presence of a 1.

When the switch is in the other position, the LED is connected to the output of the IC while the Darlington pair circuit is open. The IC is a one-shot multivibrator which serves as a pulse stretcher. In this mode, a pulse too short in duration to see with the naked eye is fed to the one-shot. It turns on for about .5 seconds enabling the LED to be seen. This circuit will respond to an input pulse width of less than 30 nanoseconds. A train of pulses causes the LED to flicker or become dimmer as the frequency of the pulses increase.

Construction Hints

An old RF probe is ideal for the case of the LOCO probe or you can make your own from tubing or even a small plastic medicine bottle. Three holes should be cut in the case for the LED, switch and power leads as shown in Figure 3-21b. The probe gets its power from the +5 volts and ground of the equipment that is being tested. The

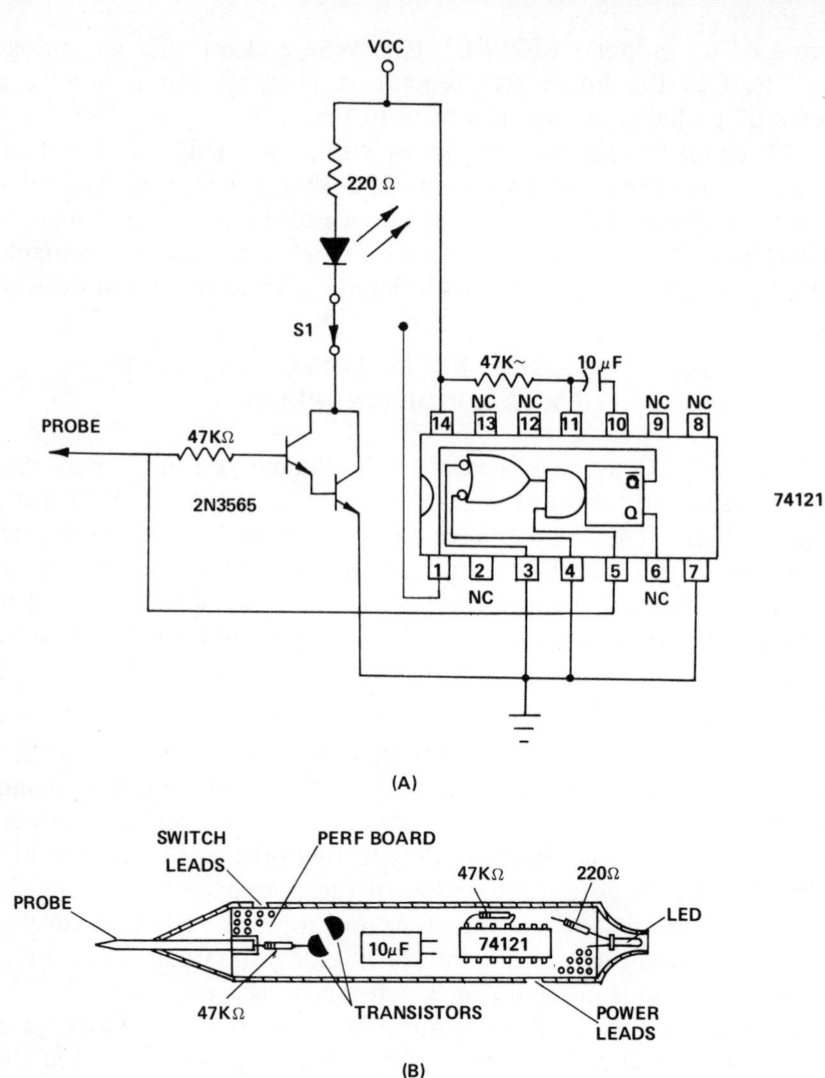

Figure 3-21. LOCO Probe (a) Schematic Diagram (b) Component Layout

power leads have mini-alligator clips attached to them. If the switch is too large to mount inside of the probe case, it can be mounted on the outside of the case as shown in Figure 3-22.

The perfboard upon which the components are mounted has .04-inch diameter holes spaced 0.100 inch on centers. This size perfboard permits the IC pins to pass through the board so that wires can

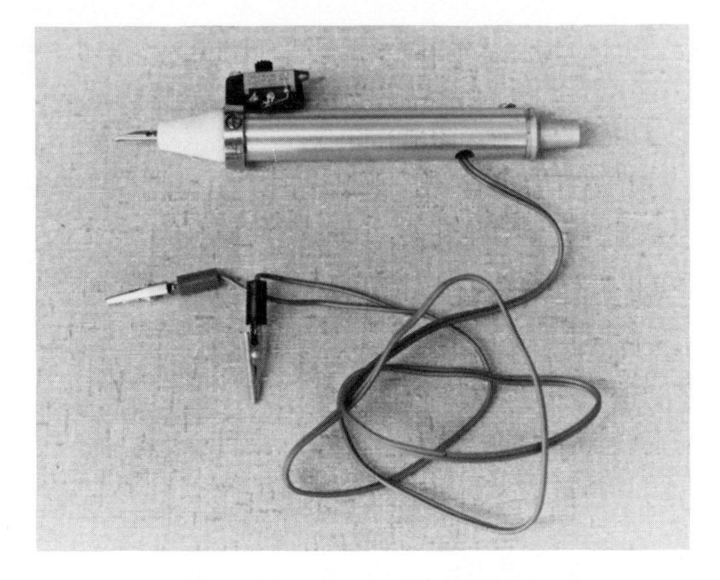

Figure 3-22. Completed "LOCO" probe

Figure 3-23. "LOCO" probe in use

be soldered to them. The probe tip can be an ordinary finishing nail wired to the perfboard and soldered into the circuit.

The circuit is first wired on the perfboard, remembering to pass the switch and power leads through their respective holes in the case before soldering is done. Then slide the perfboard very carefully into the case. Normally it is a tight fit and some of the components may have to be bent slightly. The LOCO probe is now ready for use and is shown being used with the IC breadboard/tester in Figure 3-23.

Parts List for Project #3

1—Light emitting diode (1.6-2.5 volt)
2—2N3565 Transistors or equivalent
1—74121 Integrated Circuit
1—220 ohm ¼-watt carbon resistor
2—47K ohm ¼-watt carbon resistor
1—10μf @ 10V electrolytic capacitor
1—S.P.D.T. Mini-slide switch
1—Old RF probe or small plastic medicine bottle for case
2—Alligator clips (color coded, one red and one black)
1—18-inch long double conductor of at least #22-size stranded wire for power leads

CHAPTER 4

Using Multivibrators and Flip-Flops to Produce Pulses and Store Data

An electronic oscillator changes direct current to alternating current, thus producing a frequency. Multivibrators are oscillators that primarily produce square waves and pulses. The three main types of multivibrators are:

> *Astable or free-running* where there is no stable state and the device switches back and forth between two states providing a square edge signal output. Clocks used in computers and digital circuits are astable multivibrators.
>
> *Monostable or one-shot* where there is only one stable state, from which it can be triggered to change to the other state for a predetermined time, after which it returns to the original state. Delaying and reshaping pulses is accomplished with a monostable multivibrator.
>
> *Bistable or flip-flop* where there are two stable states controlled by input signals, where it will remain in one state or the other after the input signals are removed. Flip-flops are used for storing data in the form of a 1 or 0.

There are countless multivibrators that can be built from discrete electronic components and ICs. This chapter will show you how each of the three basic type multivibrators can be constructed from basic logic gates and how each with improved circuitry and technology becomes an individual integrated circuit package.

In Project #4, you will apply your knowledge of multivibrators by adding a variable clock and bounceless switches to the breadboard/tester. The breadboard/tester will now be complete and you will be able to accurately test flip-flops and other more complex digital ICs.

4-1 ASTABLE MULTIVIBRATOR (CLOCK)

The first basic clock can be constructed from two inverters as shown in Figure 4-1.

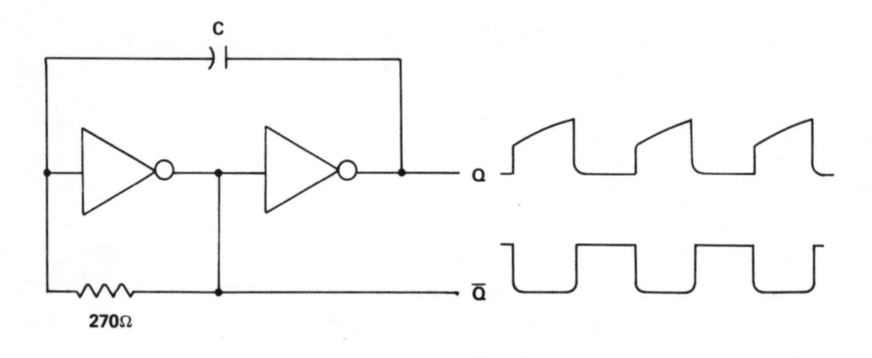

Figure 4-1. Inverter Clock

The 270 Ω resistor is used to properly bias the circuit while capacitor C provides the necessary feedback to sustain oscillation. The value of the capacitor determines the frequency of the clock, i.e.; a large capacitance produces a low frequency, while a small capacitance produces a high frequency. The approximate frequency can be found by the formula:

$$F = \frac{1}{3RC}$$

The pulses at output Q are not ideally square wave, but are satisfactory for triggering a few IC circuits. Notice that the pulses at the complementary output $\overline{Q}$ are opposite to those at output Q. When Q is high, $\overline{Q}$ is low and vice versa. Using a 7404 hex inverter IC, the output voltages should be approximately +2.8V in amplitude.

NAND Gate Clock

Another type of clock which provides sharper, and more symmetrical square waves can be constructed from a 7400 quad 2-input

NAND gate IC as shown in Figure 4-2. The pin numbers are given for each gate.

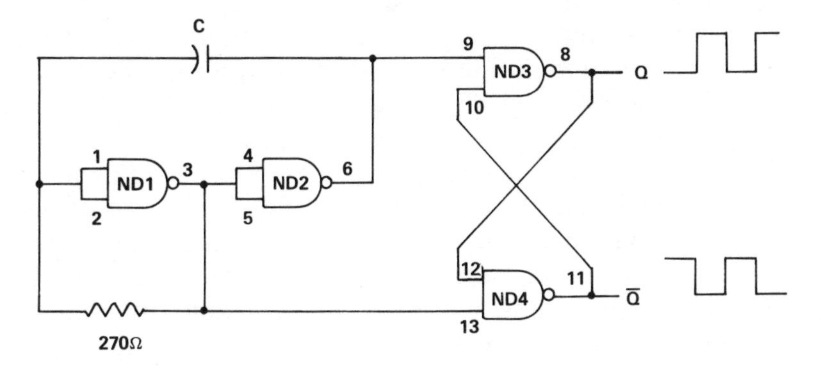

Figure 4-2. NAND Gate Clock

NAND gates ND1 and ND2 with the 270 Ω resistor and capacitor form the identical inverter clock shown in Figure 4-1. The outputs from this clock are fed into NAND gates ND3 and ND4 which form what is called a bounceless switch. When the output from ND2 goes to 0, the output from ND1 goes to 1. At this time the output from ND3 goes to 1, which is fed back into ND4 making its output go to 0. The output Q is now high while the output $\overline{Q}$ is low. Likewise, when the output from ND2 goes to 1, the output from ND1 goes to 0. Now the output from ND4 goes to 1 which is fed back into ND3 making its output go to 0. The output Q is now low while the output $\overline{Q}$ is high. This clock is better suited for triggering other ICs, since its output is more square and the voltage is approximately +3.6V in amplitude.

NOR Gate Clock

A clock which is essentially the same as the NAND gate clock can be constructed from a 7402 quad 2-input NOR gate IC as shown in Figure 4-3.

NOR gates NR1 and NR2 form the basic astable multivibrator. NR3 and NR4 provide the bounceless switch with an output of approximately +3.6V amplitude.

555 Precision Timer

The clocks mentioned thus far have some shortcomings. They are not very stable and are subject to frequency drift due to supply voltage variations, temperature change, and loading effects. The gates can only provide approximately 20 milliamperes of output current. A remarkable IC, originally developed by Signetics, is the 555 monolithic timing circuit. The 555 is an extremely stable device with a time interval variation of only .005 percent per degree centigrade temperature change. The timing accuracy of the 555 is unaffected by changes in the power supply voltage and it will operate over a range of from +5 to +18 volts. The 555 can source or sink an output current of 200 ma. The unique feature of the 555 is that it is such a versatile IC. With only a few external resistors and capacitors, the 555 can be used for precision timing, pulse generation, sequential timing, time delay circuits, pulse width modulation, pulse-position modulation, missing-pulse dectection and numerous other applications. The block diagram and pin configuration for the 555 are shown in Figure 4-4.

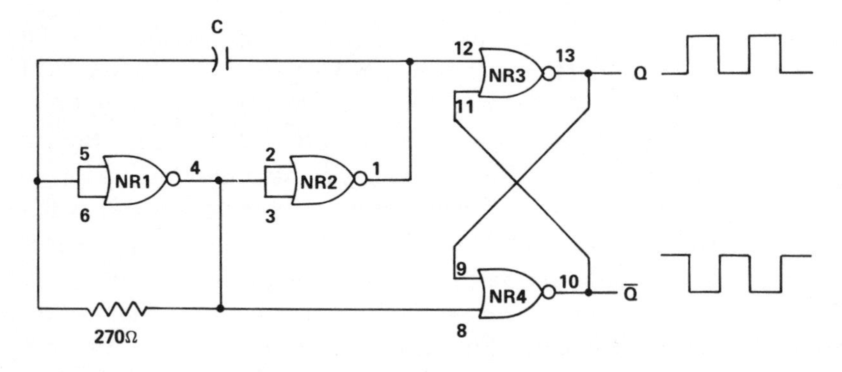

Figure 4-3. NOR Gate Clock

In the normal or quiescent state, the flip-flop biases the transistor ON which effectively places pin 7 (DISCHARGE) at ground potential. The three resistors shown form a voltage divider which sets the voltage comparator connected to pin 6 (THRESHOLD) at ⅔ of Vcc, while the voltage comparator connected to pin 2 (TRIGGER) is set at ⅓ of Vcc. The external RC time delay combination is connected between ground, pin 7, pin 6 and Vcc.

When a trigger pulse is applied to pin 2, this comparator sets the flip-flop which in turn cuts off the transistor. The external RC combination now begins it charging rate. The output stage allows pin 3 (OUTPUT) to go high during this time. When the external capacitor charges up to ⅔ of Vcc, the comparator at pin 6 resets the flip-flop which turns on the transistor, allowing the capacitor to discharge to ground. The output stage now forces the output to ground and the time delay cycle is complete. The time delay period is a function of the external RC combination and can not be changed by other trigger pulses at pin 2. However, pin 4 (RESET) can override all other inputs and can be used to initiate a new timing cycle. Pin 5 (CONTROL VOLTAGE) is used for modulating purposes or decoupling when necessary.

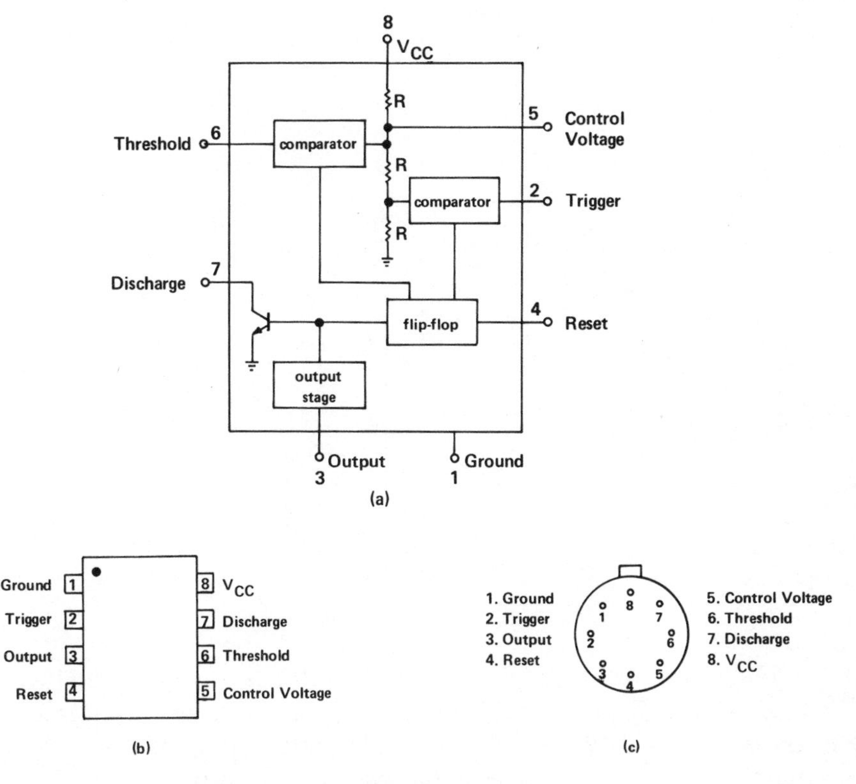

Figure 4-4. 555 Precision Timer
(a) Block Diagram (b) Dual-in-Line Package
(c) TO-5 Style Package *(Courtesy Signetics)*

555 Clock

If the 555 is connected as shown in Figure 4-5, it will trigger itself and operate as an astable multivibrator. The capacitor charges up through R_A and R_B and discharges through R_B only.

Therefore, the capacitor charges and discharges between ⅔ Vcc and ⅓ Vcc which makes the frequency independent of the supply voltage. The discharging voltage of the capacitor is applied to pin 2 which triggers the 555 and begins the charge portion of the cycle again.

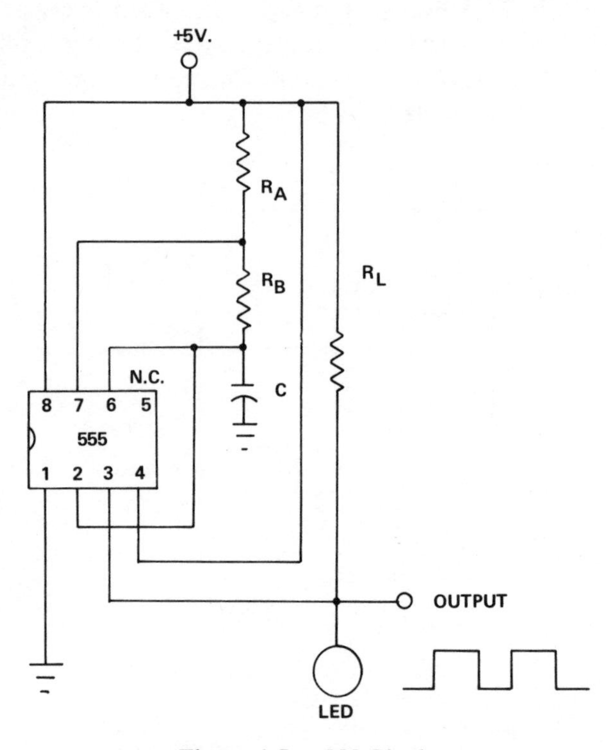

Figure 4-5. 555 Clock

The capacitor charge time when the output is high is found by $t_H = 0.693\,(R_A + R_B)C$, and the discharge time when the output is low, is found by $t_L = 0.693\,(R_B)C$. The total period is given as $T = t_H + t_L = 0.693\,(R_A + 2R_B)C$. The pulse repetition rate or frequency can then be found by

$$F = \frac{1}{T} + \frac{1.44}{(R_A + 2R_B)C}$$

The graph shown in Figure 4-6 shows how the frequency changes from 0.1 Hz to over 100K Hz when R_A, R_B and C are varied.

In order to see the LED flash at about 1.4 hertz with the circuit in Figure 4-5, let $R_A = 10\text{K}\Omega$, $R_B = 47\text{K}\Omega$ and $C = 10\,\mu\text{F}$. The load resister (R_L) can be $1\text{K}\Omega$.

It is impossible to get a symmetrical waveform (the time the output is high equals the time the output is low) because of the charge and discharge paths of the capacitor. However, the waveform can be made fairly symmetrical if R_B is approximately five times greater than R_A.

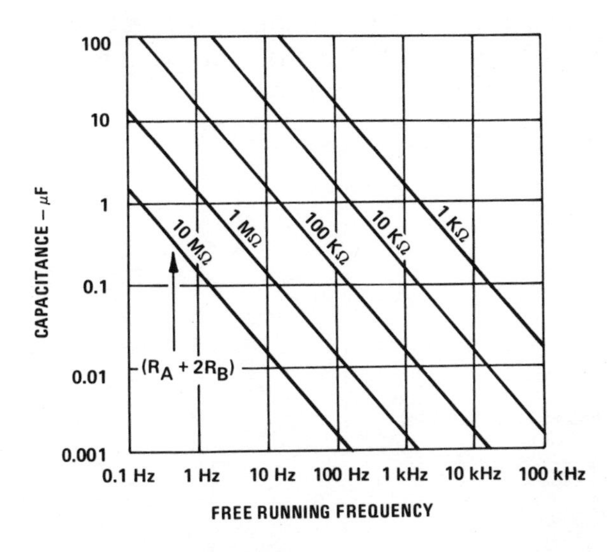

Figure 4-6. Free Running Frequency vs. R_A R_B and C *(Courtesy Intersil)**

4-2 MONOSTABLE (ONE-SHOT) MULTIVIBRATOR

As previously mentioned the one-shot (sometimes called single-shot) multivibrator can be used to reshape or delay a pulse. The time delay of a one-shot multivibrator is mainly dependent upon the time constant of a resistor-capacitor combination as shown in Figure 4-7.

The cathode of diode is connected to the 0 trigger pulse output (you will construct these trigger pulse switches in Project # 4). Normally this output is high (+3.6V). When this switch is activated, the output goes low (GND). The junction at the anode of the diode, resistor and capacitor is connected to the input of a LED to give a visual indication of the time delay. Initially the LED is on because capacitor C is charged-up to the Vcc level. When the 0 trigger pulse switch is pushed, the capacitor will discharge through the diode to ground, causing the LED to go out. When the switch is released, the capacitor will begin to charge-up exponentially. The LED will remain off depending upon the RC time constant. When the capacitor charges up to approximately +.5V, the LED driver begins to turn on, causing the LED to begin glowing. As the charging voltage across the capacitor increases, the LED becomes brighter until maximum voltage is reached. Use large values of R and C, such as R = 100KΩ and C = 1000 μf, to be able to see the time delay. Changing the values of R or C will change the time delay and control the amount of time the LED is off. This demonstration shows a negative-going time delay.

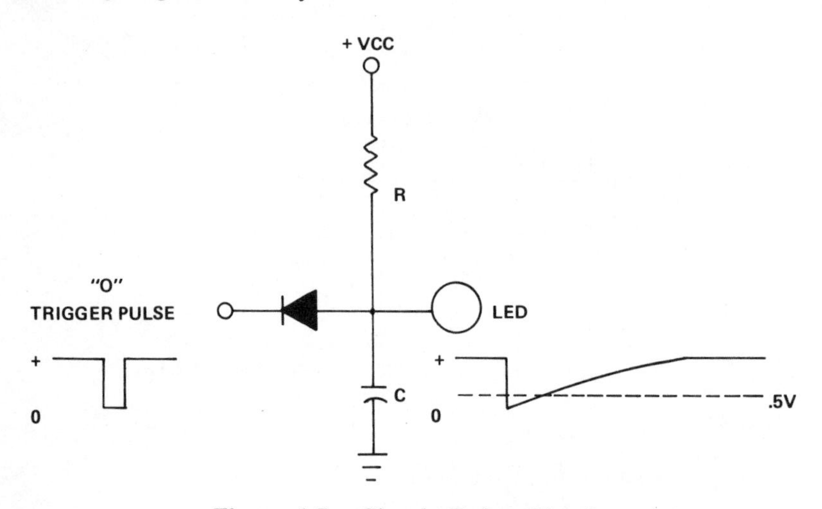

Figure 4-7. Simple Delay Circuit

NAND Gate Time Delay

A 7400 quad 2-input NAND gate IC can be constructed to produce a positive time delay with a sharp leading edge pulse as shown in Figure 4-8.

Gate ND1 acts as a control gate allowing the capacitor to discharge through the diode to ground. Gate ND2 controls the delayed output. The LEDs A and B give a visual indication of the time delay between the input pulse at the trigger and the output pulse of ND2. The 1 trigger pulse is normally low (GND). When the trigger pulse switch is pushed, a 1 (high) appears at pin 1 of ND1 causing it to turn off. LED-A, connected to the trigger pulse, is now on. The capacitor discharges through the diode and the internal structure of ND1 to ground. Pin 4 of ND2 is now 0 (low) and its output at pin 6 goes to a 1 (high) which turns on LED-B. When the switch is released, LED-A goes out, pin 1 of ND1 goes low again, turning ND1 on. The diode prevents the 1, now at pin 3 of ND1, from reaching pin 4 of ND2, thus keeping ND2 and LED-B on. The capacitor begins to charge-up toward Vcc through the internal structure of ND2. When the voltage across the capacitor charges up to the threshold voltage level for a 1 (approximately +1.3V), ND2 turns off and LED-B is extinguished.

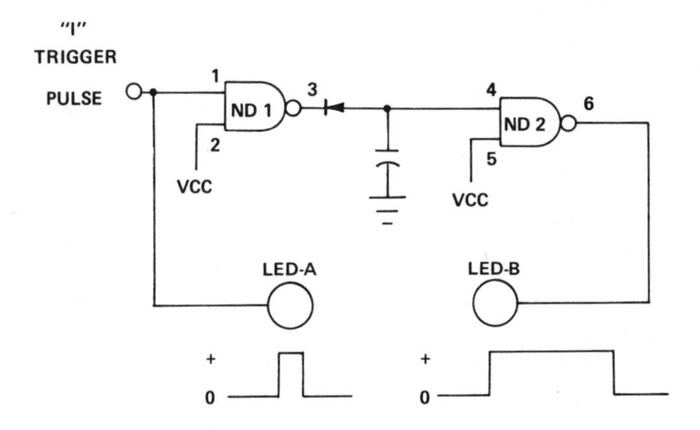

Figure 4-8. NAND Gate Time Delay

A modification of this circuit can be constructed by changing the output gate to an inverter as shown in Figure 4-9. A 7404 hex inverter IC or a 7402 quad 2-input NOR gate IC can also be used. The input of each of these circuits is connected to a 1 trigger pulse. The delayed output of each circuit is positive-going. From this figure, you can see that it is possible to use different gates to accomplish the same job in digital electronic circuits. The main consideration is to logically examine each circuit before it is constructed.

These delay circuits, of course, are not true monostable multivibrators since any input pulse occurring during the time delay period will continue to restart the time delay cycle. Also the time delay period can only be changed by changing the value of the capacitor.

555 One-Shot Multivibrator

The 555 timer makes an ideal one-shot multivibrator which is one of the reasons for its design. The circuit uses three external components; two resistors and one capacitor as shown in Figure 4-10.

A negative-going trigger pulse starts the time delay cycle which is dependent upon the values of R_A and C. If you use 100KΩ for R_A and 100 μF for C, you will be able to see approximately an eight-second time delay between LED-A and LED-B. Resistor R_L can be 1K Ω or greater for TTL operation.

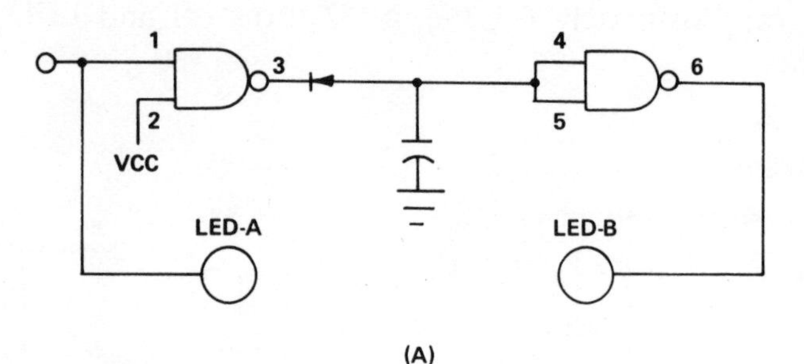

(A)

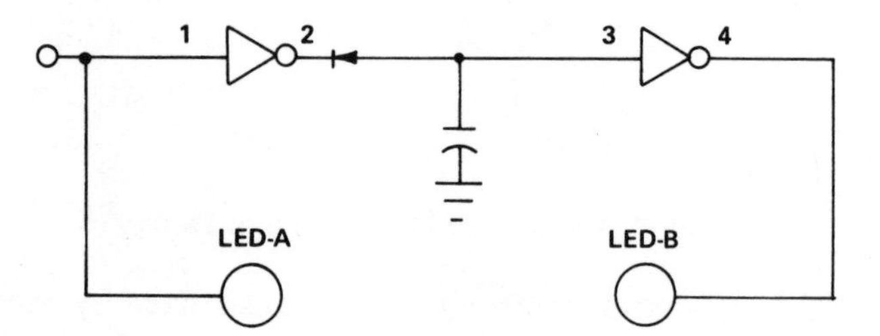

(B)

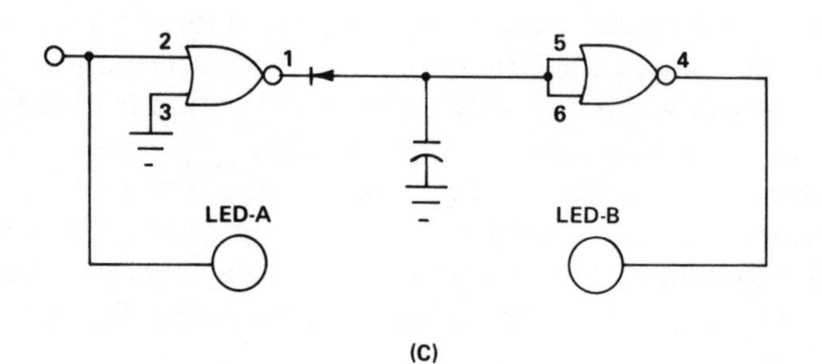

(C)

Figure 4-9. Gate Delay Circuits
(a) Using NAND Gates (b) Using Inverters
(c) Using NOR Gates

LED-A will initially be on because of the normally high condition of the "0" trigger pulse. LED-B will be off. When the "0" trigger pulse switch is pushed LED-A will extinguish and LED-B will come on. LED-B will remain on for the time delay period of R_A and

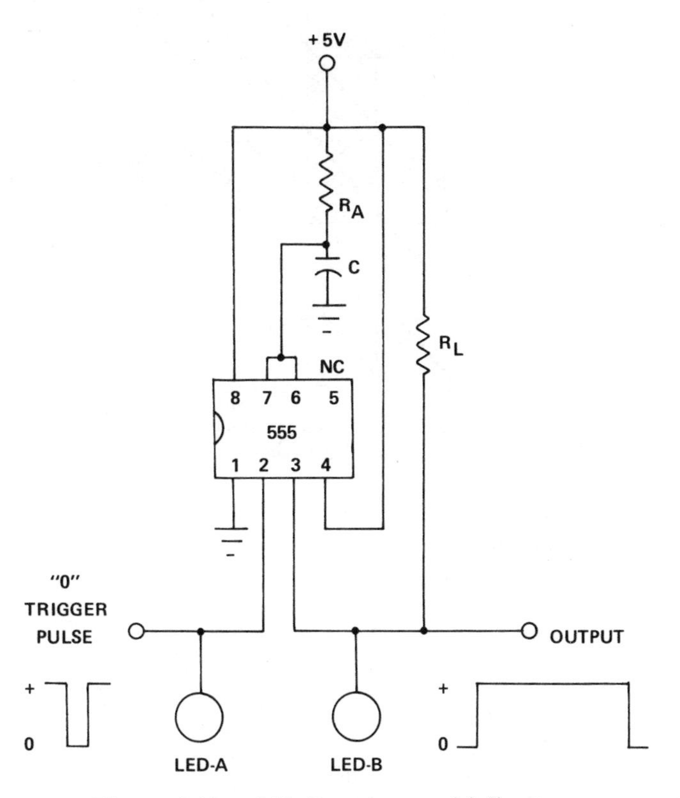

Figure 4-10. 555 One-shot multivibrator

C, after the switch is released. By pushing the trigger switch a few more times, you will see that these additional pulses do not affect the time delay period of the one-shot multivibrator. It can only be triggered on again after the time delay period has elapsed. The time delay period can be interrupted if pin 4 is disconnected from the +5v line and a negative pulse is applied to it. The time delay (when the output is high) can be found by the formula:

$$t_p = 1.1\, R_A C$$

The graph shown in Figure 4-11 gives some values of R_A and C for different time delays.

74121 Monostable Multivibrator

A special TTL one-shot multivibrator, whose time delay can be varied from 30 nanoseconds to 40 seconds, is the 74121 IC which is shown in Figure 4-12.

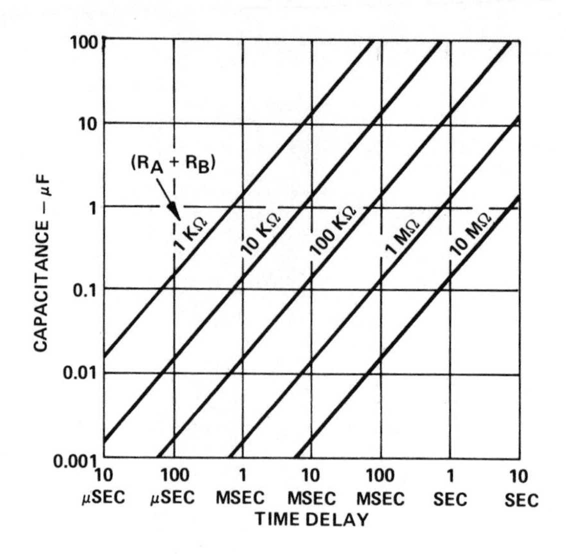

Figure 4-11. Time Delay vs. R_A, R_B and C
*(Courtesy Intersil)**

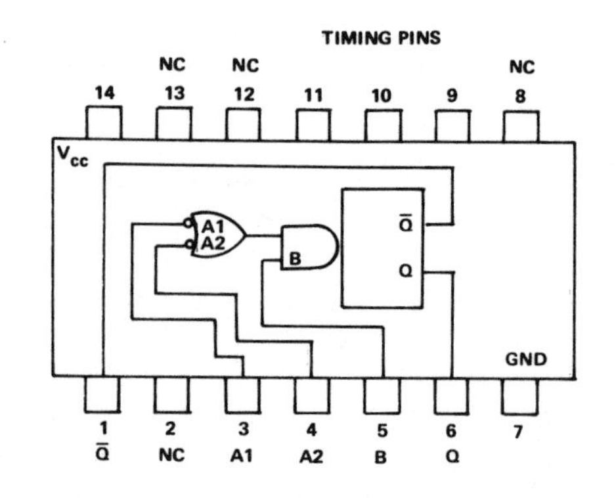

Figure 4-12. 74121 Monostable Multivibrator
*(Courtesy Signetics)**

The 74121 features positive- and negative-going triggering. When inputs A1 or A2 are low, the one-shot can be triggered by a positive pulse at input B. When input B is high, the one-shot can be triggered by a negative-going pulse at input A1 or A2. The input pulses may be of any duration, whereas the triggering occurs at a particular voltage level.

The external timing capacitor is connected between pins 10 and 11. The external timing resistor is connected from pin 11 to Vcc for an accurate repeatable pulse width. A variable pulse width can be obtained by connecting an external variable resistor between pin 9 and Vcc. Pin 9 is connected internally to a 2K Ω resistance and when pin

9 is connected directly to Vcc without an external capacitor, an output pulse width of 30 nanoseconds can be obtained. A positive Q output is available at pin 6, while a complementary $\overline{Q}$ output is available at pin 1. When pin 6 is high, pin 1 is low and vice versa. Once triggered, the outputs are independent of further transitions at the inputs and are a function only of the RC timing components.

The circuit shown in Figure 4-13 has a variable time delay from .14 to 13 seconds. Pins 3 and 4 are connected to ground and a positive trigger pulse is applied to pin 5.

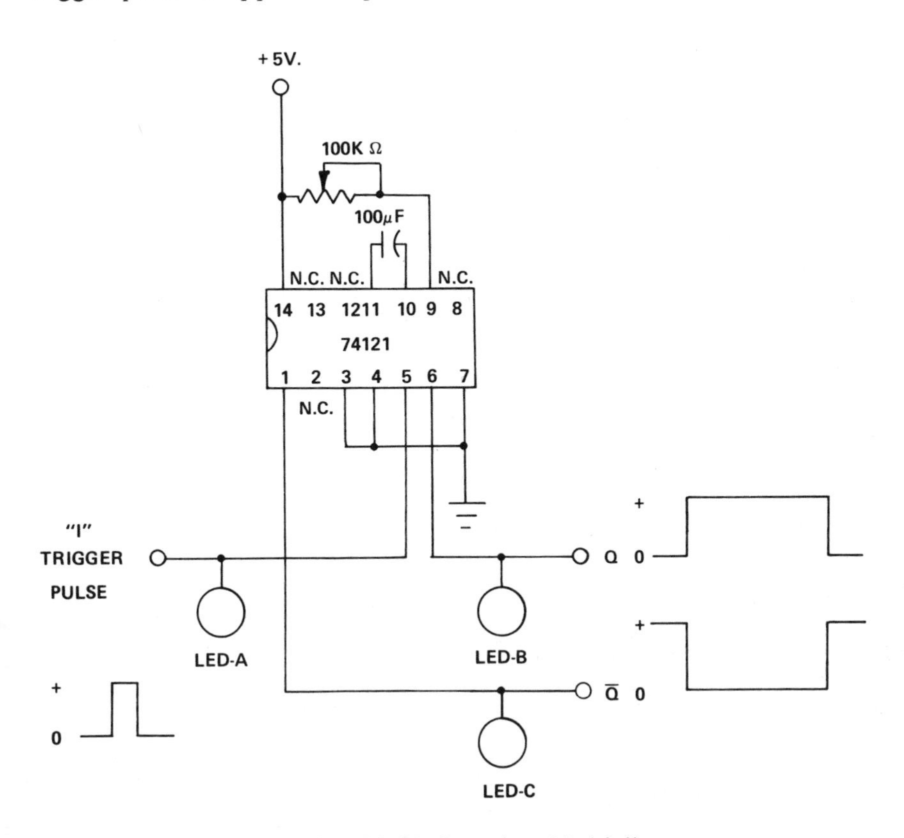

Figure 4-13. 74121 One-shot Multivibrator

An alternate method using negative-going pulses can be constructed by connecting pin 5 and either pin 3 or 4 to Vcc and connecting the other pin (either 3 or 4) to the "0" trigger pulse.

For ranges of capacitance and resistance 10pF to 10 μF and 2K Ω to 40K Ω respectively, the pulse width can be found by the formula $t_p = .693\ RC$.

4-3 SET-RESET FLIP-FLOP

The flip-flop, a fundamental memory element, is used for storing data in the form of a 1 or 0. The basic type of flip-flop is the Set-Reset flip-flop (also called the RS Flip-flop), which has two inputs and two complementary outputs as shown in Figure 4-14.

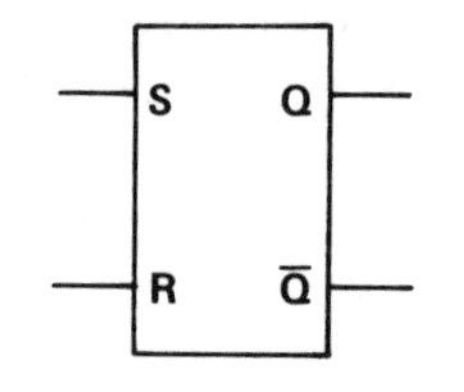

Figure 4-14. RS Flip-Flop

A pulse on the S (set) input will cause the Q output to go high, while the $\overline{Q}$ will go low. When the pulse is removed, the outputs will remain in this state. A pulse on the R (reset) input will cause the Q output to go low, while the $\overline{Q}$ output will go high. When the pulse is removed, the outputs will remain in this state. Because of this action the RS flip-flop is often called an RS latch. An RS flip-flop can be built from a basic 7400 quad 2-input NAND gate as shown in Figure 4-15. Switches A and B are input switches.

As the truth table shows, when S and R inputs are both low, the Q and $\overline{Q}$ outputs will be high, since a NAND gate only requires one input to be low for a high output. Of course, both outputs are not supposed to be in the same state for a true flip-flop. Therefore this is called a prohibited state and would normally not be used. The rest of the truth table shows that a low on either one or the other selected input (but not both) will cause the appropriate output condition and when both inputs are high the flip-flop will remain in the state previously selected. This type of RS flip-flop would require negative-going pulses to change states.

An RS flip-flop can also be built from a 7402 quad 2-input NOR gate as shown in Figure 4-16.

As the truth table shows, the prohibited state exists when both inputs are high. A high on either one or the other selected input (but not both) will cause the appropriate output condition and when both inputs are low, the flip-flop will remain in the state previously selected. This type of flip-flop requires positive-going pulses to change states.

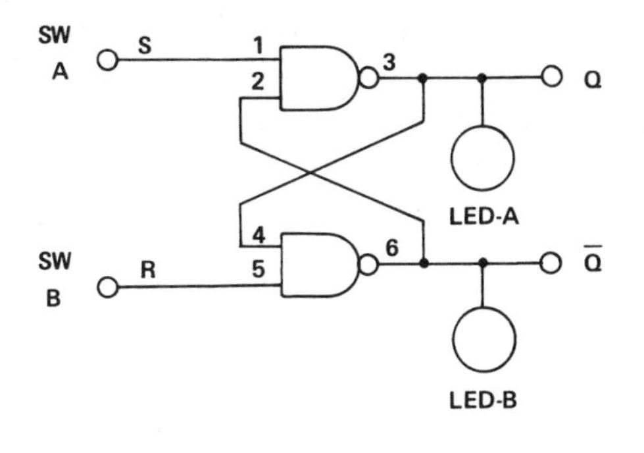

(a)

	S	R	Q	Q
*	0	0	1	1
	1	0	0	1
	1	1	0	1
	0	1	1	0
	1	1	1	0

*Prohibited

(b)

Figure 4-15. NAND Gate RS Flip-Flop
(a) Logic Diagram (b) Truth Table

Clocked RS Flip-Flop

The RS flip-flop can be modified where a third input and two control gates are added to the original circuit as shown in Figure 4-17.

The clocked RS flip-flop is more versatile and provides a control situation where two input pulses are needed to change the state of the flip-flop. As shown in the truth table, in order to turn on the flip-flop where Q = 1, a 1 must be present at the set input and the clock input (abbreviated CK, C or sometimes called Trigger or T) at the same time. In order to turn off the flip-flop where Q = 0, a 1 must be present at the reset input and the clock input at the same time. If both the S and R inputs are 0 and a 1 appears at the C input, the flip-flop remains unchanged. If both inputs are 1 and a 1 appears at the C input, the flip-flop may be in either state after the clock pulse. This situation is indeterminate and is usually avoided. This NAND gate clocked RS flip-flop responds to "1" inputs. NOR gates could also be used to construct a clocked RS flip-flop. However, this flip-flop would respond to 0 inputs instead of 1 inputs.

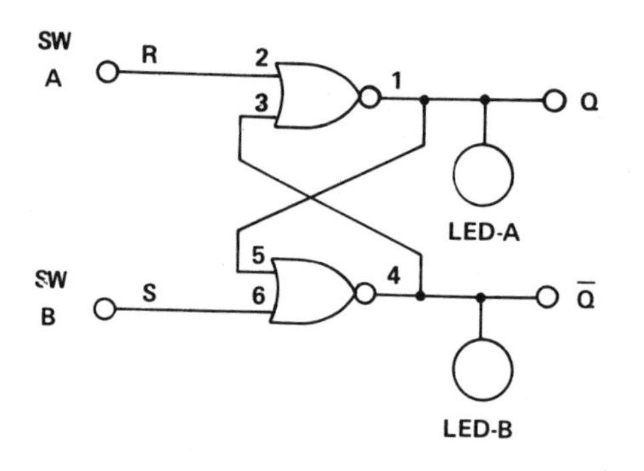

(a)

	S	R	Q	Q
*	1	1	0	0
	0	1	0	1
	0	0	0	1
	1	0	1	0
	0	0	1	0

*Prohibited

(b)

Figure 4-16. NOR Gate RS Flip-Flop (a) Logic Diagram (b) Truth Table

4-4 COMPLEMENTARY (TOGGLE) FLIP-FLOP

A complementary (also called toggle or T type) flip-flop has only one input. The T-type flip-flop will change states each time a pulse appears at the input. A 7400 quad 2-input NAND gate IC can be used to construct a T-type flip-flop as shown in Figure 4-18.

The T-type flip-flop constructed with NAND gates has one major disadvantage. Because of the transition time of the gates the trigger pulse width must be less than the propagation delay of the latch and control gates. Using a manual switch trigger pulse will most likely result in oscillations and result in indeterminate triggering. The

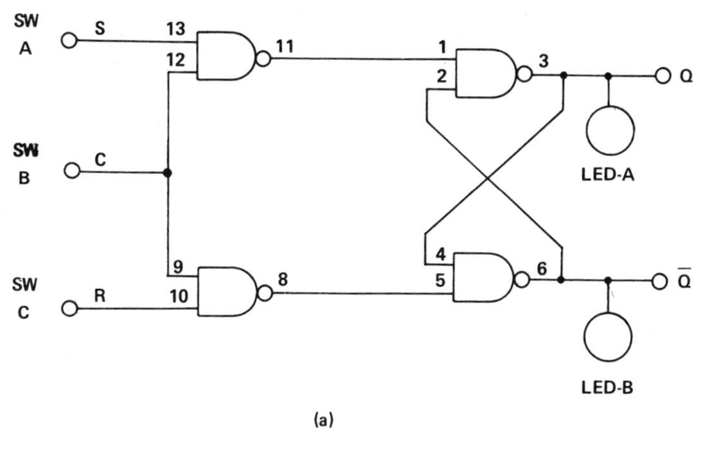

(a)

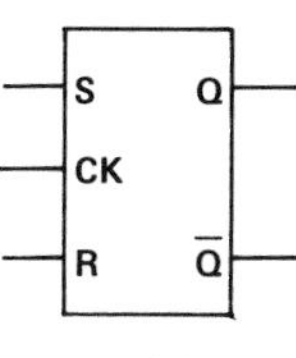

(b)

inputs		state of flip-flop after clock pulse	
S	R	Q	$\overline{Q}$
0	0	unchanged	
1	0	1	0
0	1	0	1
1	1	indeterminate	

(c)

Figure 4-17. Clocked RS Flip-Flop
(a) Using a 7400 IC
(b) Logic Symbol (c) Truth Table

flip-flop will appear not to change states. One way to overcome this problem is to use a one-shot multivibrator as shown in Figure 4-18d. The 74121 IC is wired to produce a 30 ns wide trigger pulse. The "1" trigger pulse is connected to the input of the one-shot and the output of the one-shot is connected to the T input of the flip-flop. The trigger pulse will not be visible with LED-A; however, LED-B and LED-C will change states with each trigger pulse. The truth table shows Qn + 1 and $\overline{Q}$n + 1 which refer to the state of each output after each trigger pulse.

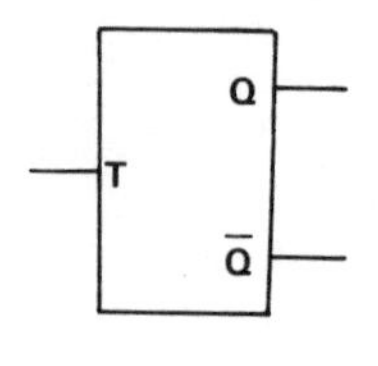

(a)

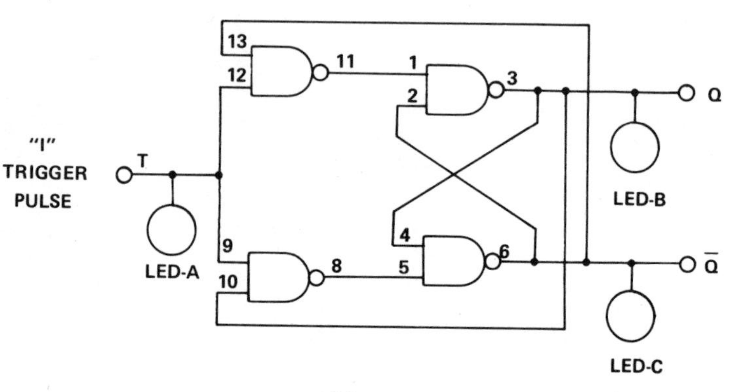

(b)

T	QN + 1	$\overline{QN}$ + 1
0	0	1
1	1	0
0	1	0
1	0	1

(c)

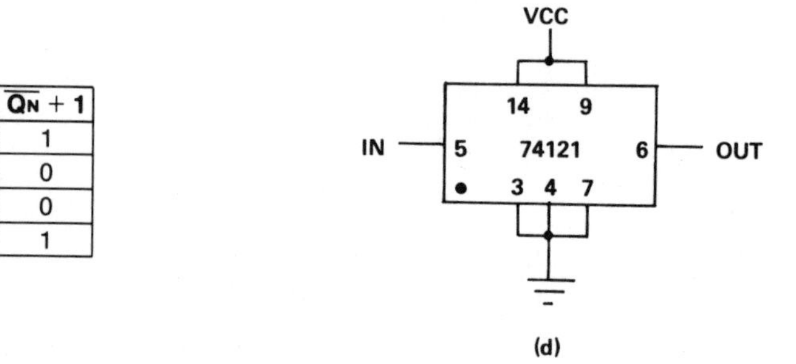

(d)

Figure 4-18. Toggle (T) Flip-Flop (a) Logic Symbol (b) 7400 IC Construction (c) Truth Table (d) One-shot

Notice the wiring diagram of the 74121 IC. The pins not used on the IC are omitted from the diagram and the other pins are rearranged to make it clearer and more readable. Most IC diagrams use this arrangement.

4-5 J-K FLIP-FLOP

The J-K flip-flop combines the features of the clocked RS flip-flop and the T-type flip-flop. The J input replaces the S input, the K input replaces the R input and the C input remains the same as shown in Figure 4-19.

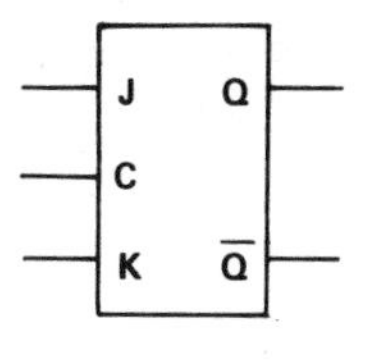

(a)

J	K	$Q_N + 1$	$\overline{Q_N} + 1$
0	0	Q_N	$\overline{Q_N}$
1	0	1	0
0	1	0	1
1	1	$\overline{Q_N}$	Q_N

(b)

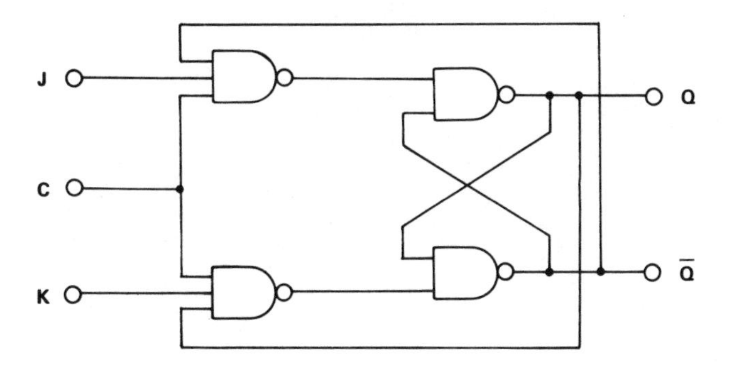

(c)

Figure 4-19. J-K Flip-Flop (a) Logic Symbol (b) Truth Table (c) Using NAND Gates

Like the clocked RS flip-flop, the J-K flip-flop also needs two input pulses in order to change states as shown by the truth table. The toggle condition exists when both J and K inputs are 1 and a clock pulse is present. The $\overline{Q}n$ in the Qn + 1 column and the Qn in the $\overline{Q}n$ + 1 column of the truth table indicate the complementing of the outputs. When both J and K inputs are 0 and a clock pulse is present, the outputs will remain in the same state. The JK flip-flop has no indeterminate condition.

A J-K flip-flop can be constructed from NAND gates as shown in Figure 4-19c. Again, a very narrow clock pulse is needed to operate the flip-flop because of the transition time of the gates. Also, two ICs would be needed to construct this flip-flop, i.e.; a 7400 quad 2-input NAND gate and a 7410 triple 3-input NAND gate.

Manufacturers of ICs have overcome these disadvantages by designing flip-flops into a single package as shown in Figure 4-20a.

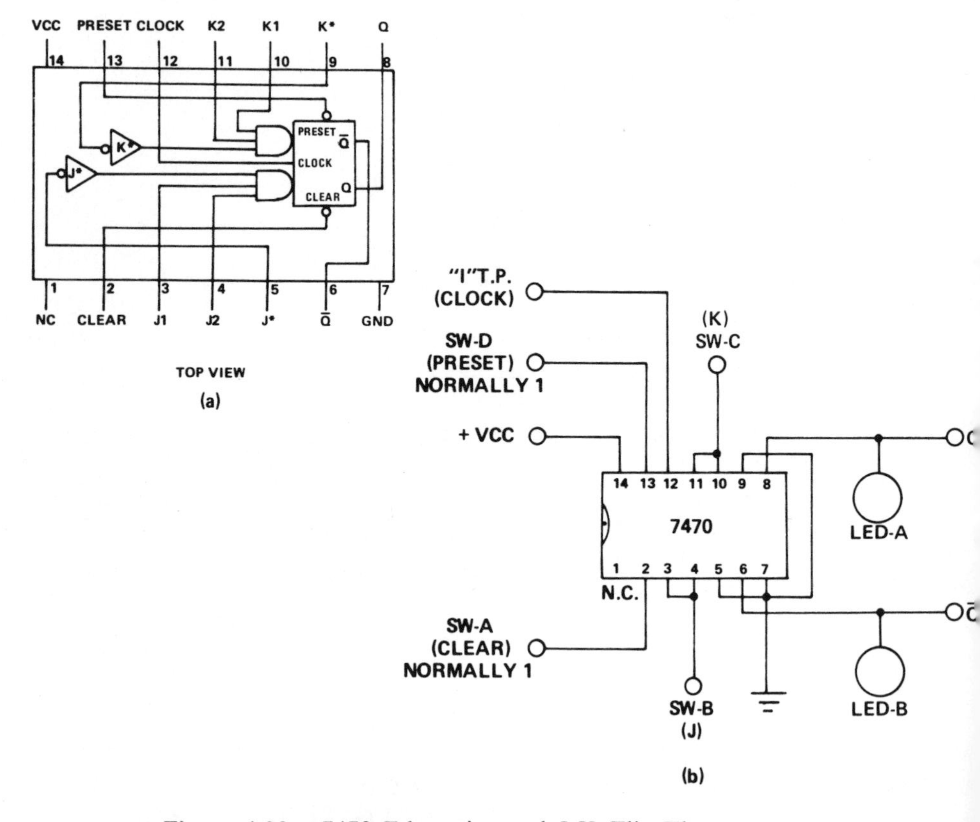

Figure 4-20. 7470 Edge-triggered J-K Flip-Flop
(a) Logic Diagram *(Courtesy National Semiconductor Corp.)*
(b) Wiring Diagram

The 7470 IC is an edge-triggered J-K flip-flop featuring gated inputs, direct clear and preset inputs. The information present at the input gates will be transferred to the outputs on the positive-going edge of the clock pulse. The multiple input gates provide more control conditions for operating the flip-flop. A low on the preset input will set the Q output to a 1, whereas, a low on the clear input sets the Q output to a 0. The clock must be low for these operations to occur. The preset and clear inputs take precedence over the other inputs. The small circles or bubbles at the clear and preset inputs to the flip-flop indicate that a low is needed to initiate any action by the flip-flop.

The 7470 can be wired to function as a single J and K input flip-flop as shown in Figure 4-20b. The clock input is connected to the 1 Trigger Pulse ("1" T.P.).

4-6 J-K MASTER/SLAVE FLIP-FLOP

Flip-flops are connected together to form binary counters (which will be studied in the next chapter.) When these counters are operated at high frequencies, the flip-flops' outputs may change too rapidly for the inputs and cause the pulses to race uncontrolled along the counter. The result is a wrong answer or count. This race problem was solved with the use of a master/slave flip-flop. Essentially, the master flip-flop is set with the positive-going edge of the clock pulse and then the information is transferred to the slave flip-flop on the negative-going edge of the clock pulse as shown in Figure 4-21.

The master flip-flop must first be set before the slave flip-flop is set and this time delay eliminates any uncontrolled pulses that may lead to an unwanted change in the input to another flip-flop.

The J-K master/slave flip-flop IC is usually not symbolically designated and can only be recognized by its number and title as shown in Figure 4-22a.

One of the flip-flops of the 7473 Dual J-K Master/Slave Flip-Flop can be wired as shown in Figure 4-22b. The same truth table used for the edge-triggered flip-flop is also used for the master/slave flip-flop. When the "1" trigger pulse switch is pushed, the outputs will not change state until it is released.

4-7 DELAY-(D-) TYPE FLIP-FLOP

The Delay- or D-type flip-flop has an output that is a function of the input that appeared one clock pulse earlier. The D flip-flop can be constructed from a J-K flip-flop and an inverter as shown in Figure 4-23.

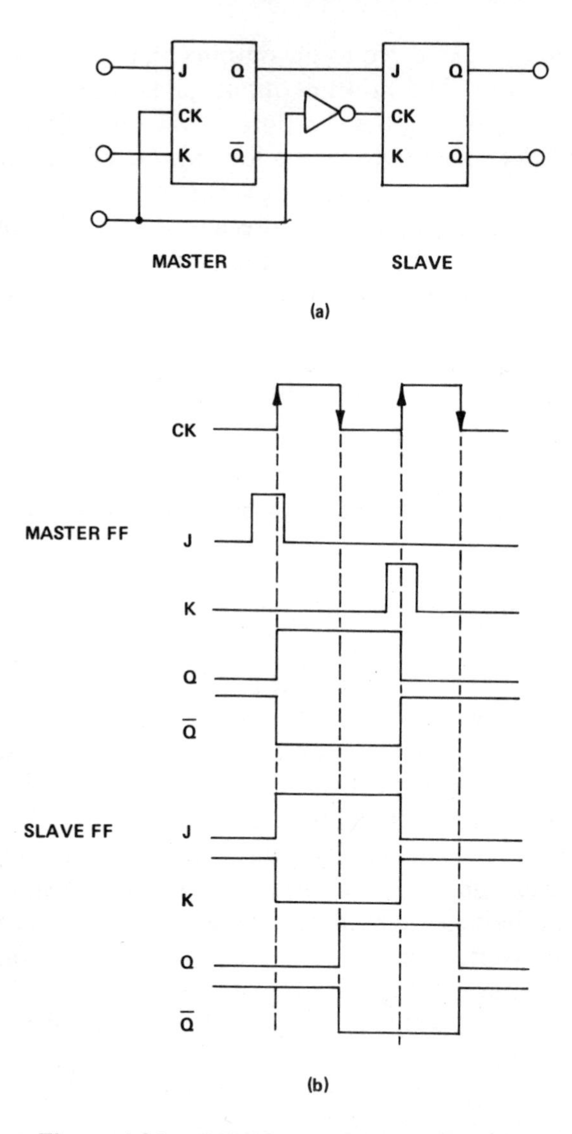

Figure 4-21. J-K Master/Slave Flip-Flop (a) logic diagram (b) Timing Chart

When a 1 is present at the D (data) input during a clock pulse, the Q output will go to a 1. When a 0 is present at the D input during a clock pulse, the inverter places a 1 on the K input and the Q output goes to a 0.

The D flip-flop is fabricated into a 7474 Dual D-Type Edge-Triggered Flip-Flop IC as shown in Figure 4-24a.

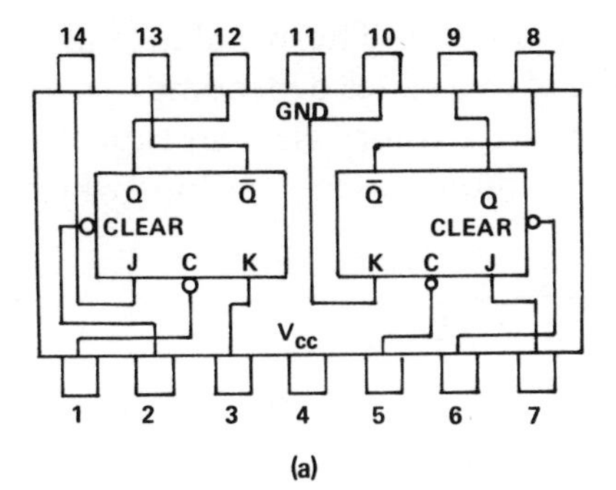

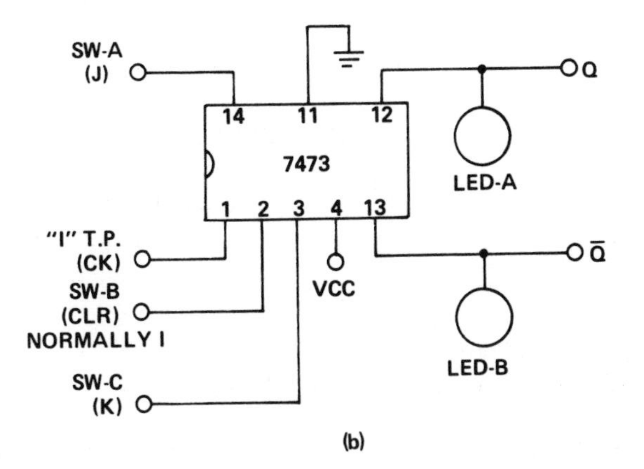

Figure 4-22. 7473 Dual J-K Master/Slave Flip-Flop
(a) Logic Symbol *(Courtesy Signetics)*
(b) Wiring Diagram for ½ IC

The 7474 IC also has preset and clear inputs and the wiring for one of the flip-flops is shown in Figure 4-24b.

SUMMARY

Multivibrators are oscillators that produce square waves which exist in two states, 1 and 0. The astable or free-running multivibrator oscillates between the two states and is referred to as a clock in digital circuits. The monostable or one-shot multivibrator once triggered will produce an output pulse whose width is dependent upon an RC time constant and then returns to the initial condition. The monostable multivibrator is used to delay, reshape and/or eliminate mechanical switch bounce. The bistable multivibrator or flip-flop will remain in one of two states depending upon the input conditions. The flip-flop is a basic memory element capable of storing a 1 and 0. There are a few commonly used flip-flops which are summarized in Figure 4-25.

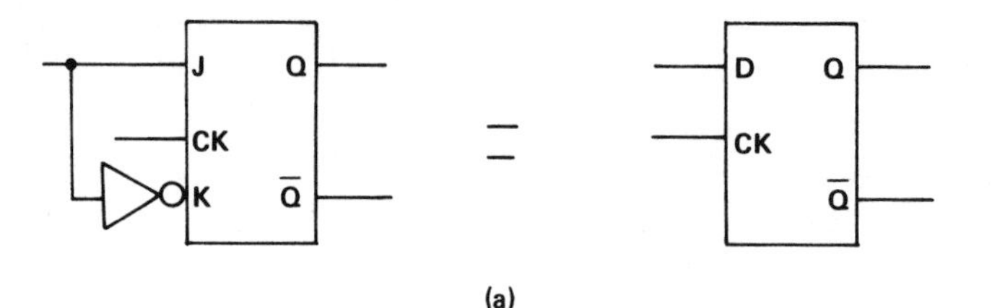

(a)

D	$Q_N + 1$	$Q_N + 1$
1	1	0
0	0	1

(b)

Figure 4-23. D Flip-Flop
(a) Logic Symbol (b) Truth Table

The RS flip-flop (a) requires one pulse to turn it on and one pulse to turn it off. With the addition of gates, the clocked RS flip-flop (b) requires two simultaneous input pulses to turn it on and off or it can be shown as a single logic symbol (c). By feeding back the $\overline{Q}$ output to the S input and the Q output to the R input, the clocked RS flip-flop forms a T flip-flop (d) whose outputs are complemented by a single input pulse and can also be shown by a more simple logic symbol (e). Combining the clocked RS flip-flop and the feedback technique produces the J-K flip-flop (f) which has both features of the previous two flip-flops. The widely used J-K flip-flop is usually shown by a more simple logic symbol (g). Adding an inverter to the K input of a J-K flip-flop produces a D flip-flop (h) where the output is the same as the input one clock pulse later. This Delay flip-flop can be shown by a more simple logic symbol (i). These flip-flops may also have clear and/or preset inputs (j) that take precedence over the other inputs.

On scratch paper, see if you can accurately write the truth table for each of these flip-flops. If you know all of the basic logic gates

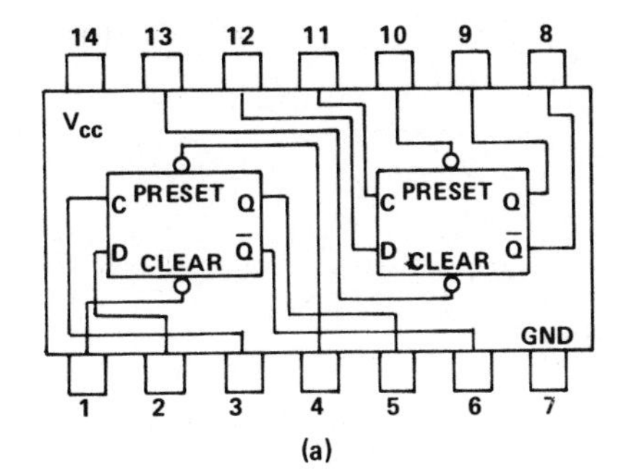

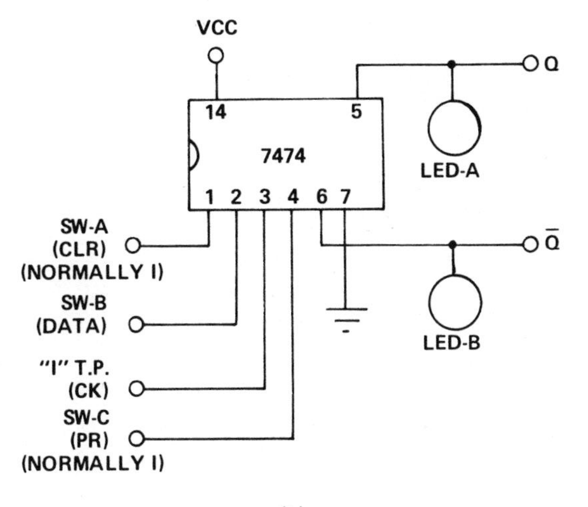

Figure 4-24. 7474 Dual D-type Edge-triggered Flip-Flop
(a) Logic Diagram (*Courtesy Signetics*)
(b) Wiring Diagram for ½ IC

and how these flip-flops work, you already have a good foundation in digital electronic circuits. These are the elements that are used to construct more useful and practical circuits that will be explained in the following chapters of this book.

4-8 PROJECT #4: BUILDING A VARIABLE CLOCK, BOUNCELESS SWITCHES AND TESTING MULTIVIBRATORS

In this project you will assemble on the breadboard/tester, the circuits used to trigger multivibrators and a clock used in digital circuits. These circuits are constructed on the lower left-hand side of the perfboard below the power supply.

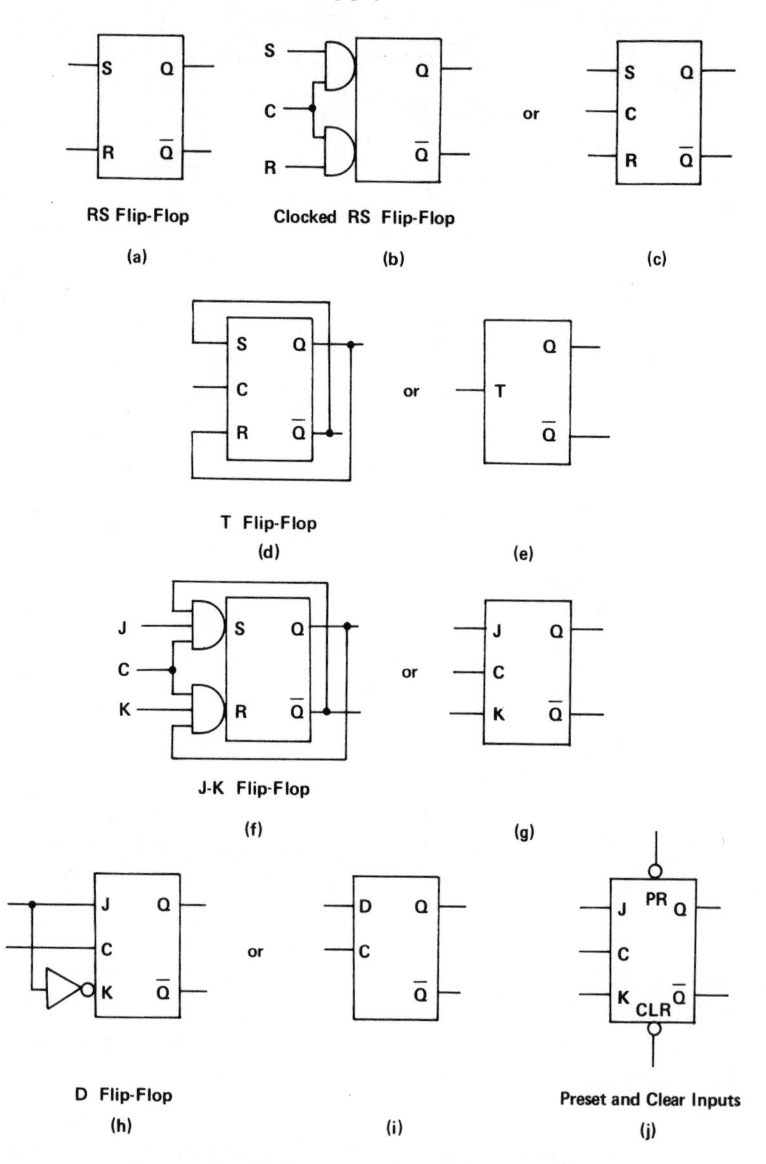

Figure 4-25. Summary of Flip-Flops

555 Variable Clock

The schematic diagram for the variable clock is shown in Figure 4-26.

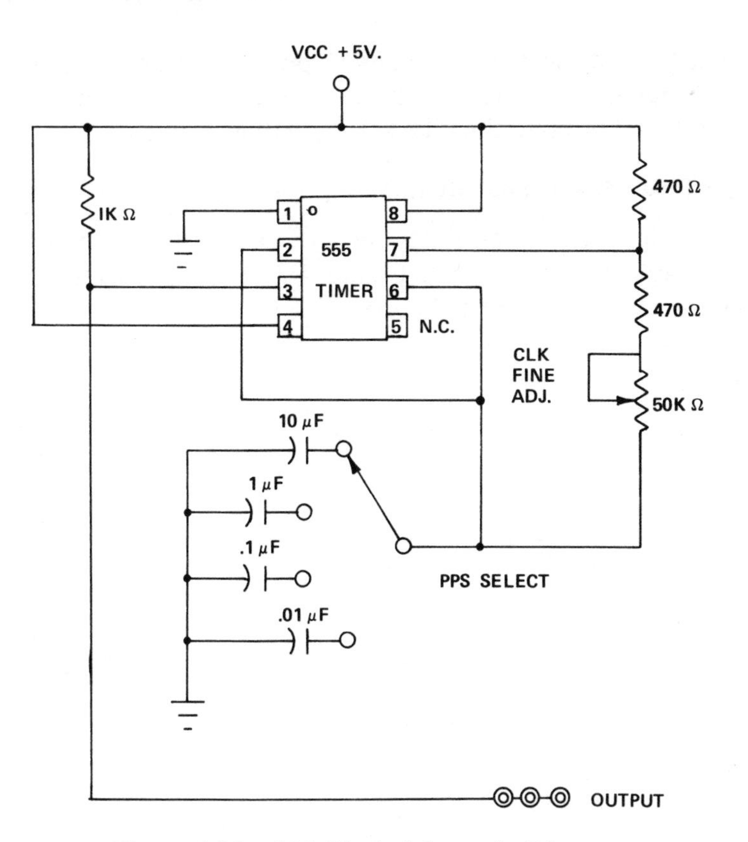

Figure 4-26. 555 Clock Schematic Diagram

All of the components are mounted on top of the perfboard, except the PPS (Pulses Per Second) switch and the Clock Fine Adjust control which are mounted through the perfboard. There are no special construction techniques needed to assemble the clock except to make good solder connections and properly dress the wires so that they lie close to the underside of the perfboard. A regular 14-pin DIP socket is used for the 555, but only the upper 4 pins on each side are wired.

The PPS switch is the range selector that selects the value of capacitance used for each range of frequencies. The approximate range of frequencies for each value of capacitance are:

10 μ f = 1-70 hertz
1 μ f = 10-700 hertz
.1 μ f = 100-7K hertz
.01 μ f = 1K-70K hertz

The Clock Fine Adjust is a 50KΩ potentiometer that adjusts the clock output to the desired frequency. The clock output terminals are located directly above the Clock Fine Adjust control.

Trigger Pulse Bounceless Switches

The trigger pulse bounceless switches are RS latches constructed from a 7400 IC as shown in Figure 4-27 and are located directly below the clock on the perfboard.

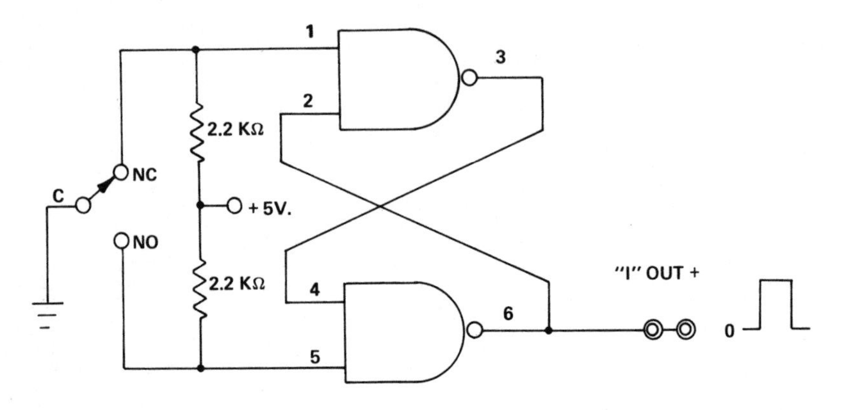

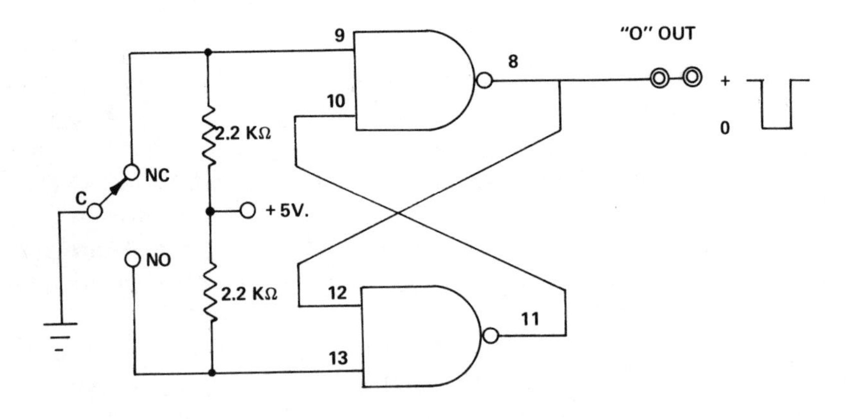

Figure 4-27. Trigger Pulse Bounceless Switches (a) "1" Output (b) "0" Output

The contacts of mechanical switches usually bounce and cause a series of minute unwanted pulses when they are operated. By using NAND gates as a latch as shown, these unwanted pulses are eliminated and a sharp leading edge pulse is generated. The pull-up resistors allow the respective inputs to the NAND gates to go to a 1 whenever the switch is not shorting that input to ground. The "1" trigger pulse will generate a positive-going pulse when operated. The "0" trigger pulse will generate a negative-going pulse when operated. The output terminals of the "1" and "0" trigger pulse are located directly above their respective momentary push-button switches.

Labels can now be attached to the perfboard for the clock and trigger pulse switches as shown in Figure 4-28 of the completed breadboard/tester.

Figure 4-28. Completed IC Breadboard/Tester

Parts List for Project # 4

2—14 pin DIP IC sockets
7—Eyelets or other type of terminals
1—Four-position mini-rotary switch with knob
2—SPDT momentary push-button mini-switches
1—555 Timer IC
1—7400 quad 2-input NAND Gate IC
1—10 μf @ 16WVDC electrolytic capacitor
1—1 μf @ 16WDC electrolytic capacitor

1—.1 μf disc capacitor
1—.01 μf disc capacitor
2—470 ohm ¼-watt carbon resistors
1—1K ohm ⅛-watt carbon resistor
4—2.2K ohm ½-watt carbon resistors
1—50K ohm linear taper potentiometer with knob

Testing Multivibrators

Now that the breadboard/tester is complete, you may proceed with the testing of multivibrators. Some resistors and capacitors are needed for the external RC time delay combinations for use with the astable and monostable multivibrators. Also, clip leads are needed to connect the external components to the IC circuits. Flip-flops can be tested by using the appropriate truth table and setting up the input switches before the clock pulse is initiated. An oscilloscope is particularly useful to observe the clock output waveforms.

Remember to handle the ICs with care and leave the power off on the breadboard/tester when installing and removing ICs from the sockets.

Table 4-1 can be used as a guide to testing the multivibrators and gives the type of multivibrator being tested, reference to its figure in the book and the ICs used.

Test # :	Multivibrator Being Tested:	Refer to Figures:	ICs Used:
1.	Inverter Clock	4-1	7404
2.	NAND Gate Clock	4-2	7400
3.	NOR Gate Clock	4-3	7402
4.	555 Clock	4-5, 4-6	555
5.	Simple Delay Circuit	4-7	NONE
6.	NAND Gate Time Delay	4-8	7400
7.	Gate Delay Circuits	4-9	7400, 7402 7404
8.	555 One-shot	4-10, 4-11	555
9.	74121 One-shot	4-12, 4-13	74121
10.	RS NAND Gate Flip-Flop	4-14, 4-15	7400
11.	RS NOR Gate Flip-Flop	4-16	7402
12.	Clocked RS Flip-Flop	4-17	7400
13.	T Flip-Flop	4-18	7400, 74121
14.	J-K Flip-Flop	4-19, 4-20	7470
15.	J-K Master-Slave Flip-Flop	4-21, 4-22	7473
16.	D Flip-Flop	4-23, 4-24	7474

Table 4-1. Multivibrator Tests

CHAPTER 5

Using Registers and Counters for Basic Digital Operations

Basic logic gates and multivibrators are combined into more complex units which are used for handling binary numbers and performing arithmetic operations. These units can be constructed from several of the ICs you have studied in the preceeding chapters. However, the 7400 TTL series has some of these complex units standardized onto a single chip. Less external wiring is needed when more circuitry can be incorporated into a single IC. This makes it easier to design and maintain IC circuitry which, in many cases, results in a more economical and efficient device.

In this chapter you will learn how logic gates and flip-flops are arranged to produce registers and counters. You will be able to construct these circuits from the experience gained in Chapters 2 and 4.

In Project # 5 you will gain experience working with the more complex ICs and be able to wire up registers and binary counters.

5-1 REGISTER OPERATIONS: STORING, SHIFTING AND TRANSFERRING

A register is a group of flip-flops used for the temporary storage of binary data. The number of flip-flops determines the amount of data per unit and is often referred to as *computer word, data word,* or just *word.* A register may have a work length of 4 bits which means it contains 4 flip-flops. A register must not only be able to store data, but also move the data in a specified operation. Some registers are designed to accomplish only a few operations. The 4-bit register shown in Figure 5-1 will perform most of the basic operations required. This register can be constructed from a couple of 7476 Dual JK Master/Slave Flip-Flop ICs.

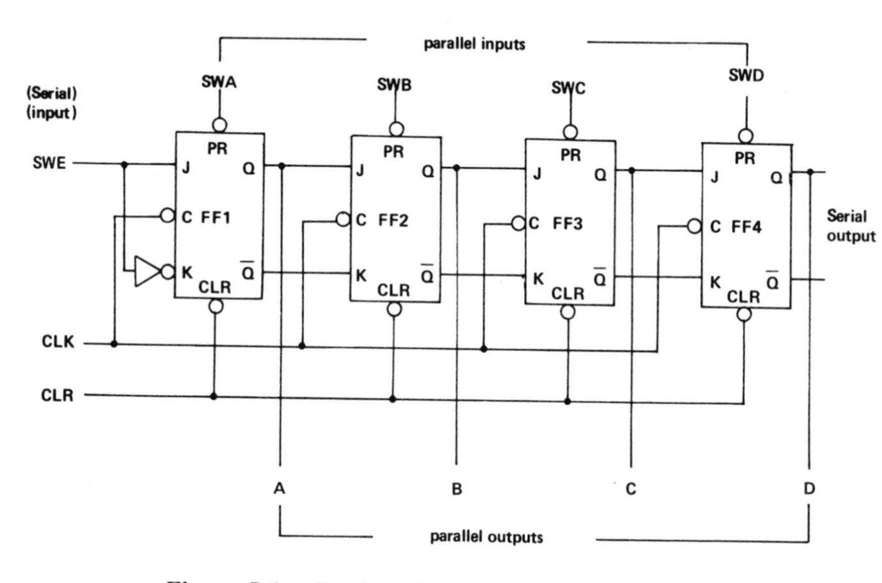

Figure 5-1. Register Storing and Shifting Right

Parallel Inputs and Parallel Outputs

Data can be entered at the parallel inputs whenever the CLR line is high and any switch A through D is in a 0 state. When no data is to be entered the switches should be high or 1. As an example, When SWA=1, SWB=0, SWC=1 and SWD=0, then FF1 is off, FF2 is on, FF3 is off and FF4 is on. When all switches are again high, the register will be storing the binary number 0101. The Q outputs will be FF1=0, FF2=1, FF3=0 and FF4=1. This data can be fed to other digital circuits via parallel outputs A through D. A 0 pulse occurring on the CLR (clear) line will turn off all of the flip-flops and their Q outputs will be 0.

Serial Input, Shifting and Serial Output

The register shown in Figure 5-1 has a normal shift right operation. The outputs of one flip-flop feed into the inputs of the next flip-flop. If one of the flip-flops has a Q output of 1, the following flip-flop has a 1 on its J input. When the next clock pulse appears, that flip-flop will turn on. Likewise, a $\overline{Q}$ output of 1 will be applied to the K input of the following flip-flop and turn it off when the next clock pulse appears. Data is allowed to enter the register at the serial input when SWE is 1 and a clock pulse occurs. The inverter at the K input of FF1 turns off FF1, when a 0 is entered at SWE. It takes 4 clock pulses to completely load the register.

Table 5-1 will demonstrate the serial loading and shifting action of the register when it contains the binary number 1010 and the condition of each flip-flop after each clock pulse:

Clock Pulse	Input SWE	Register Contents				
		FF1	FF2	FF3	FF4	
no pulse	0	0	0	0	0	
1	0	0	0	0	0	
2	1	1	0	0	0	
3	0	0	1	0	0	
4	1	1	0	1	0	← Register Loaded
5	0	0	1	0	1	
6	0	0	0	1	0	
7	0	0	0	0	1	
8	0	0	0	0	0	← Register Cleared

Table 5-1. Timing Sequence for a Shift Register

As seen from the table, after the register is loaded, 4 more pulses will shift the data completely out of the register. This data is lost unless the serial output is connected to another register or other digital circuit.

Registers such as these are of course, fabricated into one IC. Figure 5-2 shows the 7491 8-bit shift register. Inputs are on pins 11 and 12, while the outputs are on pins 13 and 14.

Shift Right or Shift Left Register

With the addition of AND-OR logic gates and inverters, the register can be reconstructed to accomplish a shift right or shift left operation as shown in Figure 5-3a.

The JK flip-flops are actually D-type flip-flops and a couple of 7474 Dual D Flip-Flop ICs could be used which would not require the use of separate inverters. Each flip-flop has an AND-OR gate combination connected to its J input, which allows data to be shifted into it from the preceeding flip-flop or from the following flip-flop. The SHR (shift right) control line and the SHL (shift left) control line determine in which direction data will be shifted. As an example, if FF3 is on, its Q output places a 1 on AND gate 4. If the SHL line is 1, then AND gate 4 and OR gate 2 allow the 1 to pass and condition the J input of FF2. When the next clock pulse occurs, FF2 will turn on completing a shift left operation. Similarly, FF3 places a 1 on AND gate 7 and if the SHR line is 1, then AND gate 7 and OR gate 4 allow the 1 to pass and condition the J input of FF4. When the next clock pulse occurs, FF4 will turn on completing a shift right operation.

The 7495 is a 4-bit right-shift/left-shift register IC as shown in Figure 5-3b and will be used in Project # 5.

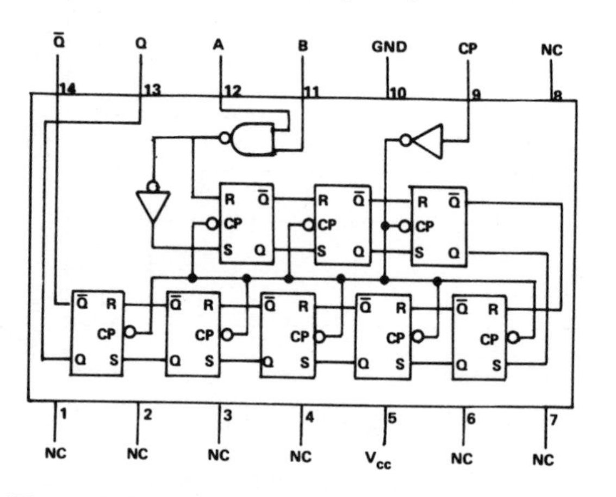

Figure 5-2. 7491 8-bit Shift Register IC
(Courtesy National Semiconductor Corporation)

Transferring

Data can be moved from one register to another by serial shifting as you have seen. The time it takes to shift a data word from one register to another depends upon the number of bits of the word and the clock frequency. This is time-consuming and the faster method of parallel transfer between digital circuits is sometimes desired. One method of parallel transfer between two registers is shown in Figure 5-4.

Many of the lines that are normally present have been omitted to show only the importance of parallel transferring. The Q and $\overline{Q}$ outputs of each flip-flop of register A are fed into a pair of NAND gates

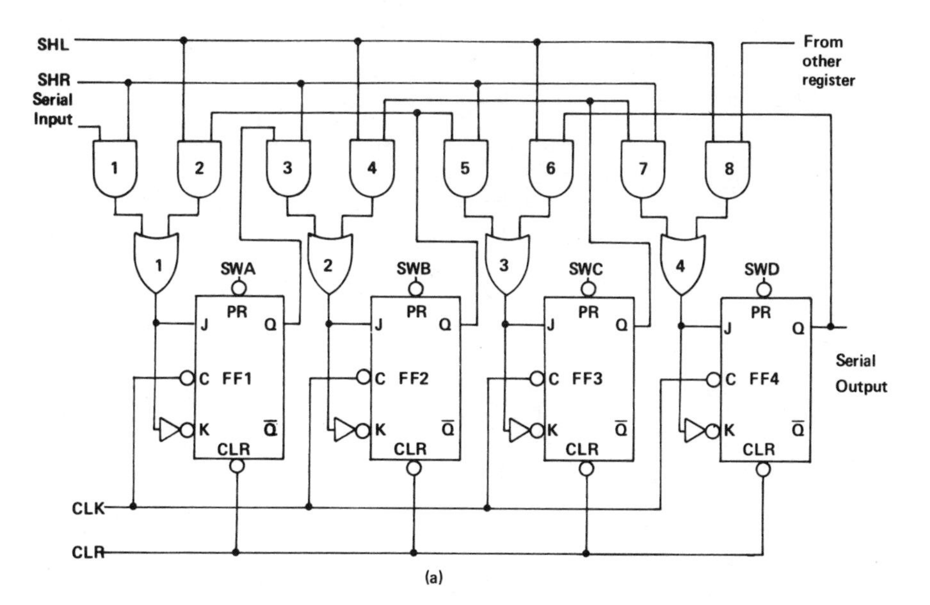

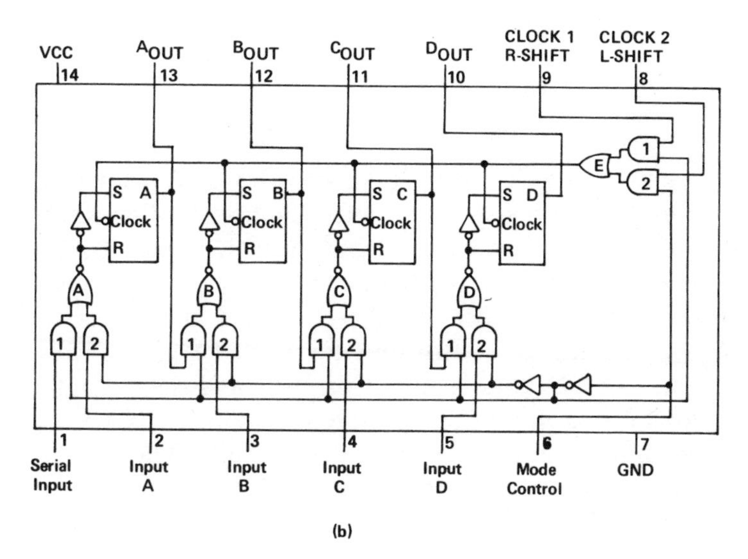

Figure 5-3. Shift Right or Shift Left Register
(a) Additional AND-OR Logic Gates (b) 7495 4-bit Right-shift/Left-shift Register IC
(Courtesy National Semiconductor Corporation)

which are connected respectively to the preset and clear inputs of each flip-flop of register B. A transfer line also connects into each NAND gate which controls the transfer operation.

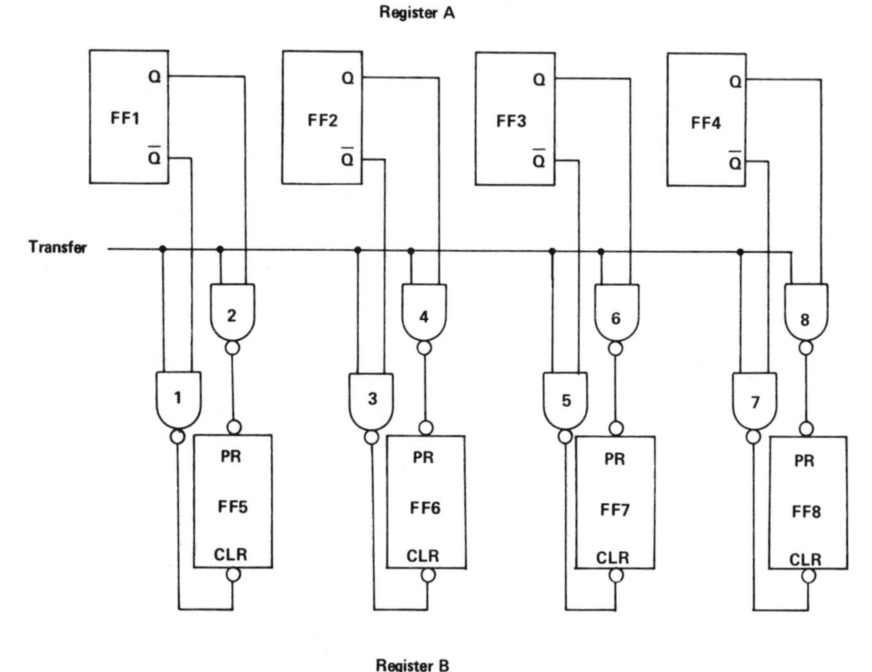

Figure 5-4. Parallel Transfer

Assume that FF1 is off, FF2 is on, FF3 is off, FF4 is on, and all of the flip-flops in register B are off. The transfer line is 0, therefore all of the NAND gates have a 1 output which doesn't allow any of the preset and clear lines to operate the flip-flops in register B. When the transfer line goes to a 1, NAND gates 4 and 8 turn off, which allows the 0 outputs to turn on FF6 and FF8 respectively. At the same time NAND gates 1 and 5 turn off and their 0 outputs keep FF5 and FF7 turned off, respectively. The other NAND gates remain on, but do not affect any other circuits. The binary number 0101 has been transferred from register A to register B. Register A still contains the number 0101, but can be cleared to accept new data for other operations.

5-2 RING COUNTER USED FOR DISTRIBUTING PULSES

The ring counter is essentially a shift register with its output coupled back into its input and can be used to distribute pulses from a clock to different circuits at various times. A ring counter using J-K flip-flops is shown in Figure 5-5.

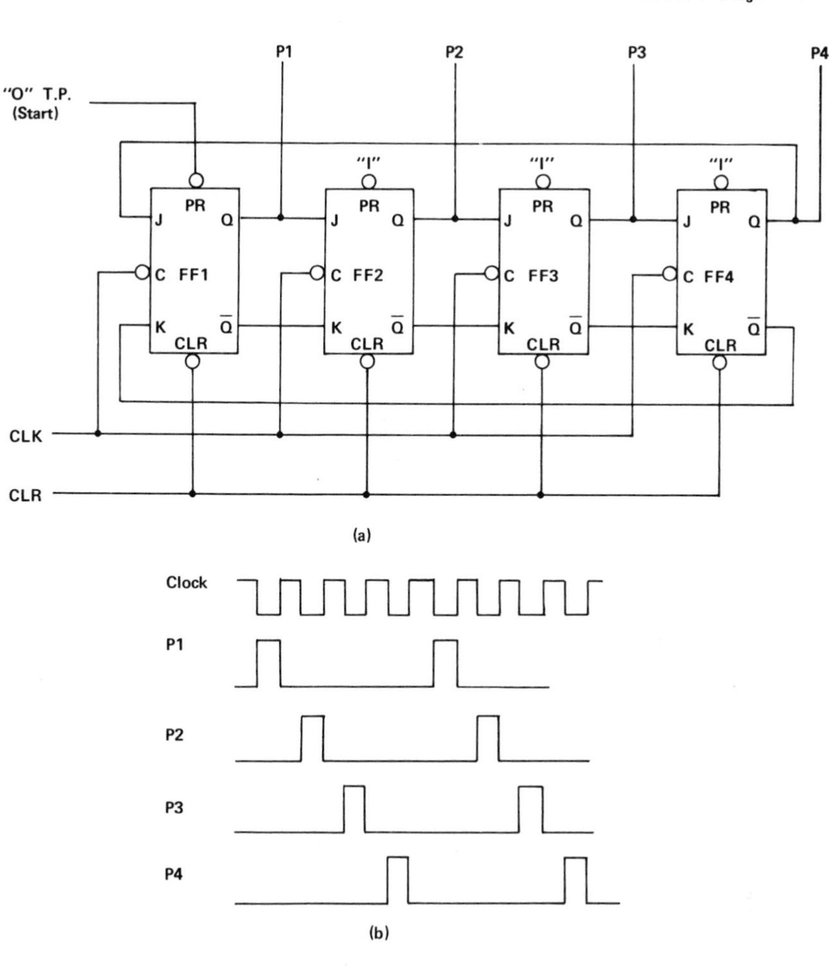

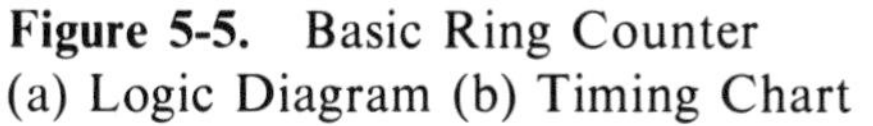

Figure 5-5. Basic Ring Counter
(a) Logic Diagram (b) Timing Chart

The ring counter is started by a 0 trigger pulse to the preset of FF1. This pulse must be shorter in duration than a clock pulse, or two or more flip-flops could turn on at the same time. The idea is to have one pulse "ripple" along the register to produce the various output pulses shown in the timing chart. If the clock is set at a high frequency, it may be necessary to use a one-shot multivibrator to trigger (start) the operation.

When FF1 is on, a 1 appears on the J input of FF2. The next clock pulse will turn on FF2. The $\overline{Q}$ output of FF4 being 1 is fed back to the K input of FF1 which turns off this flip-flop at the same time. The shifting action continues along the register until the Q output of FF4 is fed back to FF1 to initiate another cycle. The output pulses are, of course, taken off the Q outputs of each flip-flop. A clear pulse applied to all flip-flops stops the shifting action regardless of which flip-flop is on at the time.

Switch-Tail Ring Counter

The "switch-tail" ring counter is the same as the normal ring counter, except the output is cross-coupled back to the input as shown in Figure 5-6.

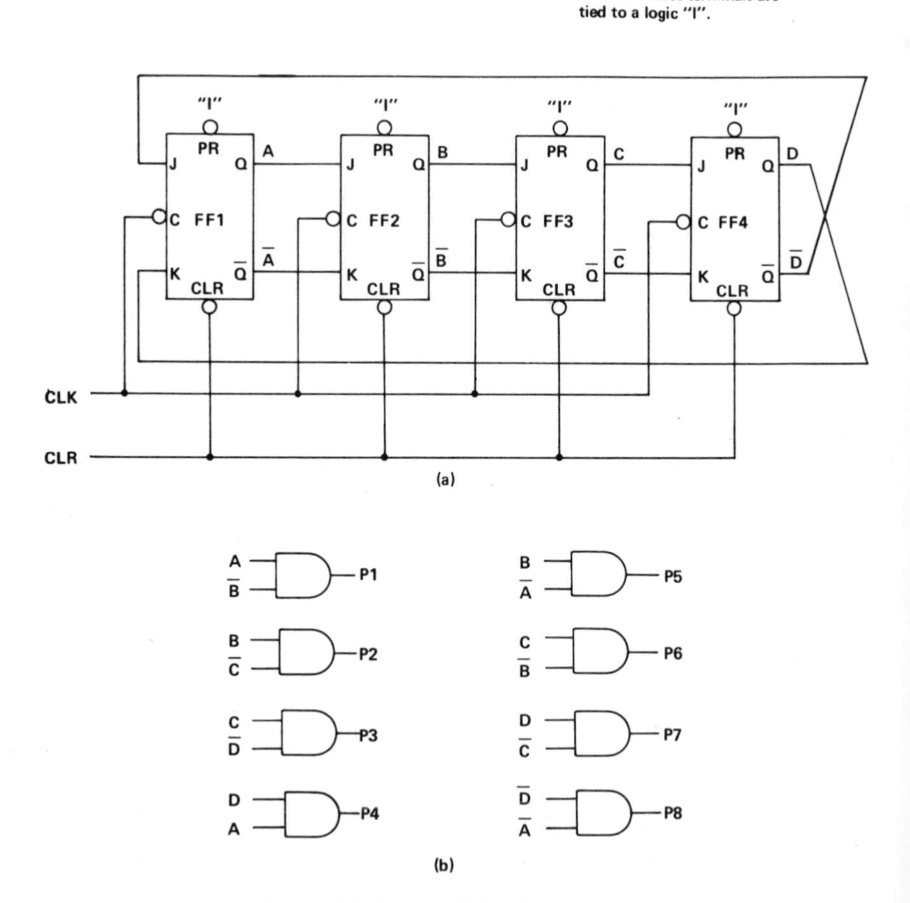

Figure 5-6. "Switch-Tail" Ring Counter
(a) Logic Diagram (b) Pulse Decoder Diagram

This counter is self-starting since the $\overline{Q}$ output of FF4 is applied to the J input of FF1. Each flip-flop will turn on in succession until all of the flip-flops are on. Then each flip-flop turns off in succession until all of the flip-flops are off and the cycle is complete.

AND gates or NAND gates can be connected to the Q and $\overline{Q}$ outputs to form a decoder that will provide 8 separate pulses, although the counter contains only 4 bits. Decoders will be explained in Chapter 7.

An interesting visual observation can be seen with these ring counters when the outputs are connected to LEDs with the clock set at a very low frequency. Most likely, you have seen this type of display in movies and television shows about computers and space adventures.

5-3 BINARY UP-COUNTER USED FOR INCREMENTAL ADDING

Binary counters are very important in digital systems. The binary counter is made up of a group of flip-flops used to tally a series of input pulses, as well as to display them in binary form, temporarily store the numbers and use them in other operations.

The basic binary up-counter is constructed from T type flip-flops which can be made from J-K flip-flops as shown in Figure 5-7. The Q and $\overline{Q}$ outputs are cross-coupled back to the J and K inputs of the same flip-flop.

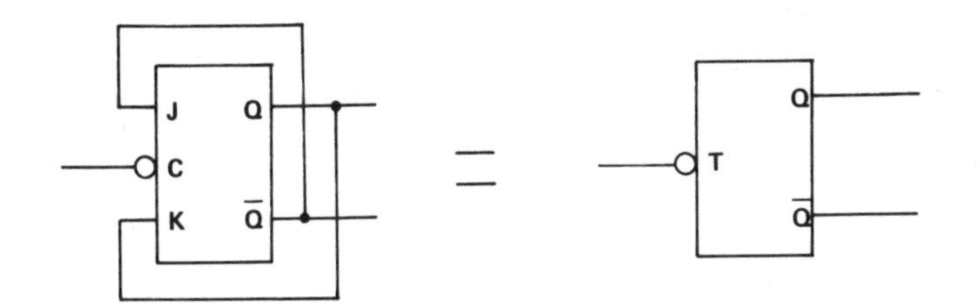

Figure 5-7. J-K Flip-Flop to T-type Flip-Flop

The TTL IC flip-flop will perform the same toggle function when its J and K inputs are simultaneously tied to a 1 or high.

A 4-bit binary up-counter is shown in Figure 5-8. The series of input pulses may come from a manual switch clock, or other device, and enters FF1 at the T input. The Q output of each flip-flop is connected to the T input of each succeeding flip-flop. This arrangement is referred to as asynchronous, since the input pulses are transferred through the counter, stage-by-stage. Because of their action, the binary counter is often referred to as a "serial" or "ripple" counter. Each flip-flop will toggle only when its trigger input goes low. The input data travels from the lease significant bit flip-flop to the most significant bit flip-flop, from left to right. The LEDs are cross-coupled to the Q outputs to permit viewing of the binary number in the normal position.

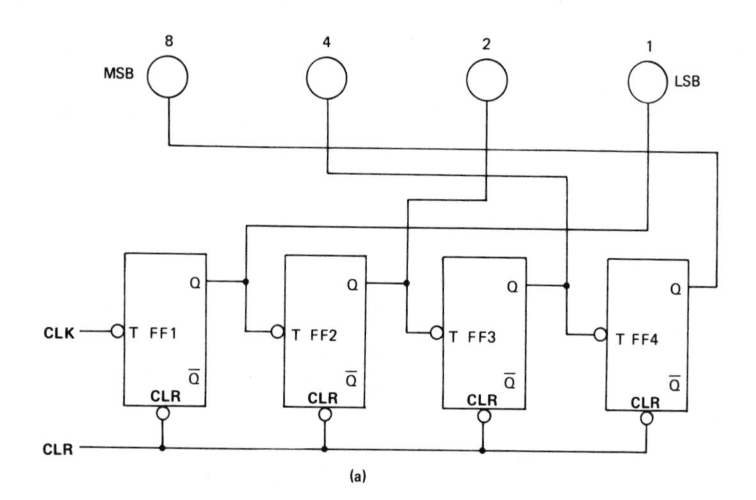

(a)

Clock Pulse	FF4	FF3	FF2	FF1
0	0	0	0	0
1	0	0	0	1
2	0	0	1	0
3	0	0	1	1
4	0	1	0	0
5	0	1	0	1
6	0	1	1	0
7	0	1	1	1
8	1	0	0	0
9	1	0	0	1
10	1	0	1	0
11	1	0	1	1
12	1	1	0	0
13	1	1	0	1
14	1	1	1	0
15	1	1	1	1

(b)

Figure 5-8. Binary Up-Counter
(a) Logic Diagram (b) Logic Table

The counting operation is simple:

Clock Pulse 1 turns on FF1. Clock pulse 2 turns off FF1. The Q output of FF1 goes low, which turns on FF2. Clock pulse 3 turns on FF1 again. Clock pulse 4 turns off FF1, which turns off FF2. The Q output of FF2 goes low and turns on FF3. This procedure is continued until 15 clock pulses have entered FF1 and all of the flip-flops are on. Clock pulse 16 turns off all of the flip-flops and the counter is said to be reset or clear. If the input pulses continue, the counter will perform the count up again. The clear line can, of course, reset the register at any time regardless of the number in the counter. The number of pulses that a counter may count is infinite depending only on the number of flip-flops it contains. With relation to binary positional notation, a counter with 4 flip-flops can count to 15, with 5 flip-flops to 31, etc.

The binary counter is also used as a frequency divider. Each flip-flop requires 2 pulses at its input to complete a cycle, therefore, the output frequency of each flip-flop is:

FF1 = 1/2 f
FF2 = 1/4 f
FF3 = 1/8 f
FF4 = 1/16 f
etc. ...., where f is the input frequency

Synchronous Binary Up-Counter

The delay that results by counting stage-by-stage with the binary counter is difficult to interface with other digital circuits using a clocked system. In high frequency systems this problem can very easily cause errors. A synchronous counter as shown in Figure 5-9 can minimize this problem.

The essential requirement of a synchronous counter is that the clock input of each flip-flop is driven by the system clock at the same time. This allows all conditioned flip-flops to change state simultaneously, synchronous with the clock. The 74163 uses inhibit gates which control the clock pulse to each flip-flop. During a count operation, the J and K inputs of all flip-flops are held high by conditioned NAND gates. The clock pulse is prevented from reaching a flip-flop unless the low order flip-flops before it are on. For example; Q_C will go high when the next clock pulse occurs if Q_A and Q_B are high. The 74163 also has gates for controlling the count, parallel loading and clearing operations.

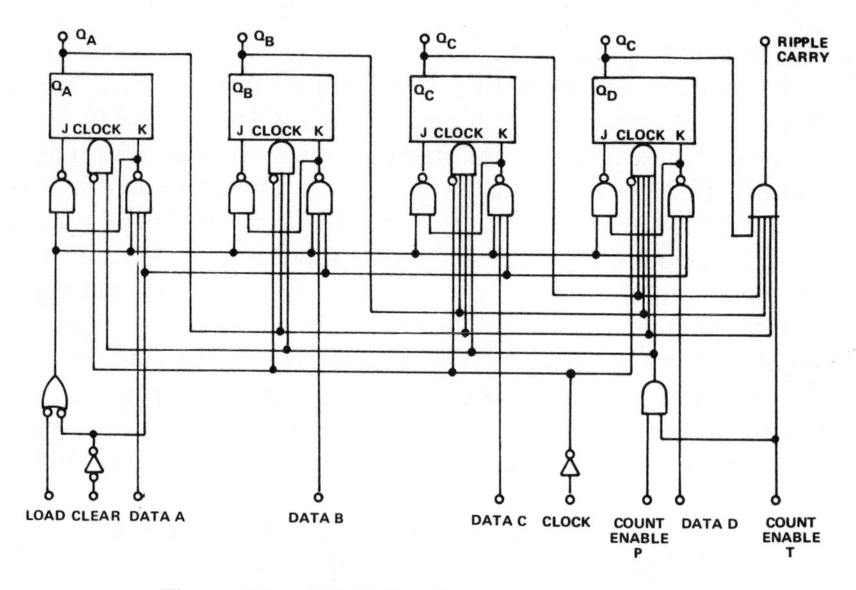

Figure 5-9. 74163 Synchronous 4-bit Counter
(Courtesy Signetics)

5-4 BINARY DOWN-COUNTER USED FOR INCREMENTAL SUBTRACTING

The binary down-counter operates similar to the binary up-counter except its count is reduced by one with each clock pulse. The down-counter could be used in a basic subtracting operation or perhaps used as an indicator when a certain set of sequences has been completed. The down-counter is identical to the up-counter, except the $\overline{Q}$ outputs are fed to the T inputs as shown in Figure 5-10.

Its counting operation is also simple:

Clock pulse 1 turns FF1 on. Its $\overline{Q}$ output goes low, which turns on FF2. Likewise, FF3 and FF4 also turn on and the counter contains the binary 15 number. Clock pulse 2 turns off FF1. Its $\overline{Q}$ output goes high and nothing else occurs in the counter. Clock pulse 3 turns FF1 on again and its $\overline{Q}$ output goes low, turning off FF2. Clock pulse 4 turns off FF1 and nothing else occurs in the counter. Clock pulse 5 turns on FF1 and FF2, which turns off FF3. This procedure continues until clock pulse 16 has entered FF1 and all of the flip-flops are off indicating that the counter is at zero. If the input pulses continue, the counter will perform the count-down again.

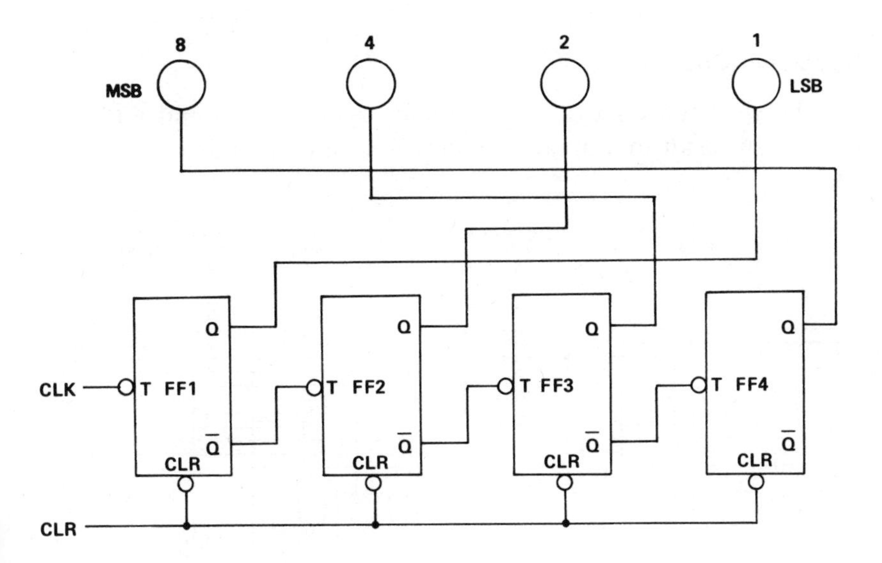

(a)

Clock Pulse	FF4	FF3	FF2	FF1
0	0	0	0	0
1	1	1	1	1
2	1	1	1	0
3	1	1	0	1
4	1	1	0	0
5	1	0	1	1
6	1	0	1	0
7	1	0	0	1
8	1	0	0	0
9	0	1	1	1
10	0	1	1	0
11	0	1	0	0
12	0	1	0	0
13	0	0	1	1
14	0	0	1	0
15	0	0	0	1

(b)

Figure 5-10. Binary Down-Counter (a) Logic Diagram (b) Timing Chart

Up/Down-Counter

Down-counters can also be made synchronous and added circuitry can result in a binary up/down-counter as shown in Figure 5-11.

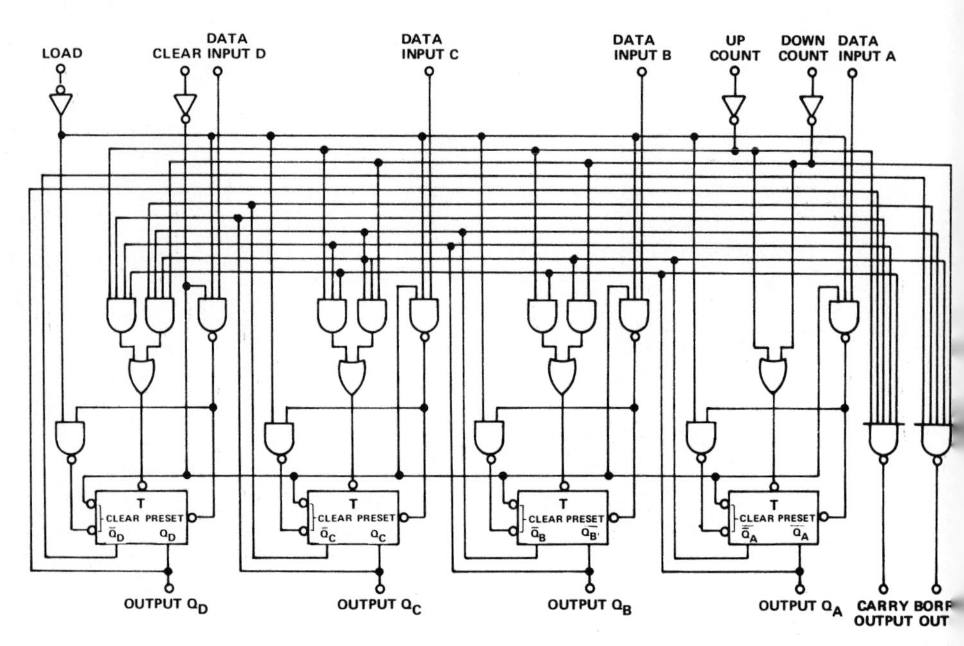

Figure 5-11. 74193 Synchronous 4-bit Binary Up/Down Counter *(Courtesy Signetics)*

The direction of counting is determined by which count input is "0" pulsed while the other count input is held high. Parallel data inputs allow desired data to be preset in the counter when the load input is low, which also allows the counter to function as a modulo-N counter. An extra clear input is provided, which takes precedence over other operations when a high level is applied. A carry output and borrow output are provided for arithmetic operations. This IC is very versatile when used in a complex digital system.

5-5 MODULO COUNTER USED FOR INTERMEDIATE COUNTING

A 3-bit counter can count up to 7 and a 4-bit counter can count up to 15. But, what if you need a counter to count to 10, reset and count again? You can use a 4-bit counter that can be wired in various ways to accomplish this. A counter that is used to count specific amounts is referred to as a modulo counter or simply a MOD-counter. A MOD-10 counter will reset on the 10th input pulse, therefore, the counter will have accepted 9 input pulses before resetting.

Feedback Method

A MOD counter can be constructed using the feedback method as shown in Figure 5-12.

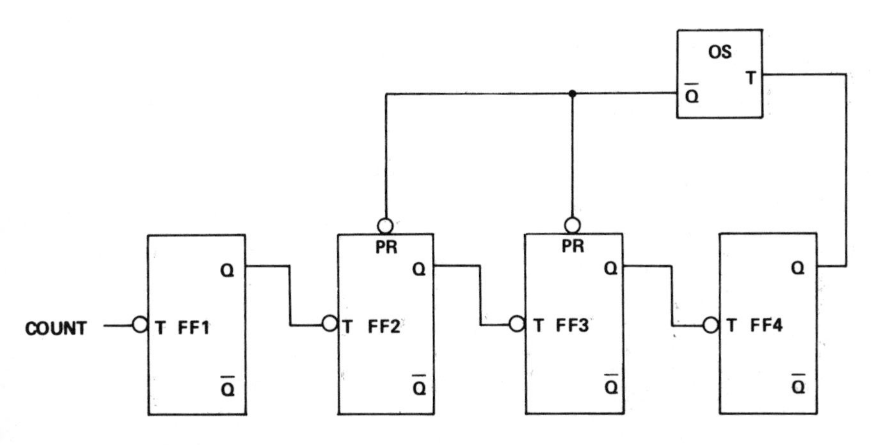

Figure 5-12. MOD-10 Counter Using Feedback

The Q output of FF4 is fed back to the preset inputs of FF2 and FF3 via the one-shot multivibrator. The counter begins counting in the normal manner. When pulse 8 turns FF4 on, the Q output goes high, which turns on FF2 and FF3, thereby advancing the contents of the counter to 14. The 9th pulse turns on FF1 and the counter is full. The 10th pulse will then reset the counter. This sequence will repeat if input pulses are present. The one-shot multivibrator provides the "0" pulse from its $\overline{Q}$ output to preset FF2 and FF3 and then goes high again to allow all of the flip-flops to reset on the 10th pulse.

Reset Control Method

Another method of constructing MOD-counters is with reset control as shown in Figure 5-13.

The Q outputs of FF2 and FF4 are fed to a NAND gate which

in turn is connected to the counter reset line. The counter begins counting normally. When the counter reaches the binary number 10, the Q outputs of FF2 and FF4 turn off the NAND gate which resets the counter to zero. This sequence will continue as long as input pulses are present.

A single 7493 4-bit binary counter IC is capable of constructing from a MOD-2 to a MOD-16 counter. The 7490 decade counter IC and the 7492 divide-by-12 counter IC are also MOD-counters. These counters are given in Project #5.

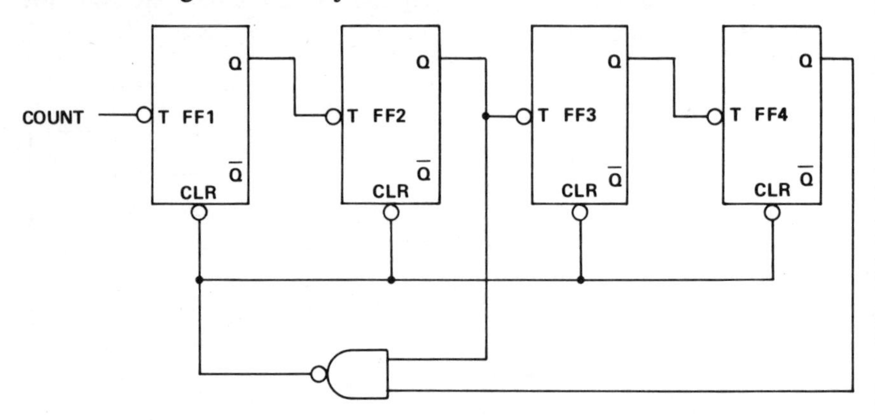

Figure 5-13. MOD-10 Counter with Reset Control

SUMMARY

You have seen how flip-flops are combined into registers which are used for temporarily storing data. These registers have the capability of shifting and transferring the data to other registers or digital circuits. A shift register with its output fed back to the input produces a ring counter which can be used as a pulse distributor. The same ring counter with its output cross-coupled back to the input produces the "switch-tail" ring counter and with a decoder network provides eight separate pulses.

Binary ripple counters are made from T-type flip-flops and can be used for keeping track of a number of events or occurences. These counters can be made to count up, to count down, or count up to some specified number before resetting and are called Modulo counters. Counters can be made synchronous where each flip-flop is preconditioned and all flip-flops operate in unison with the clock pulse. This not only increases the operating speed of the counter, but reduces the chance of errors due to the race problem.

5-6 PROJECT # 5: CONSTRUCTING AND TESTING DIGITAL REGISTERS AND COUNTERS

In Project #5 you will gain experience in reading and wiring more complex digital circuits. The IC breadboard/tester is used to wire-up and test up to four ICs. If you desire to design and construct circuits which require more than four ICs, another perfboard with mounted IC sockets could be connected to the breadboard/tester.

There are various ways in which IC wiring diagrams are drawn. This project uses a straight wiring procedure on some, showing all pins on the ICs in their normal positions. Other drawings will have pins rearranged on the ICs and those pins not used will be omitted to simplify the circuit and make it easier to read. You should still orient all ICs in the same direction and will have to locate the proper pins by counting them. Remember, also, the difference in numbering between a 14 pin and 16 pin IC.

Turn off the power on the breadboard/tester when you wire-up, rewire or disassemble the IC circuits.

5-6A Register Storing and Shifting

In this section, you will construct a standard 4-bit shift register from two 7476 ICs as shown in Figure 5-14. You will be able to leave the standard shift register assembled and perform the operations of Section 5-6B, the ring counter and Section 5-6C, the switch-tail ring counter by simply adding two more wires. All of the clock inputs to the flip-flops are tied together and can be manually clocked using the "0" trigger pulse switch or the regular clock output set to your desired shifting speed.

Switches A-D are connected to the preset input of the flip-flops respectively, which allows you to load the desired data into the register. Switch E allows you to enter the data in a serial manner as each clock pulse occurs. Switch F, which is set normally high, allows you to clear the register at any time by moving it to the low position.

Switch G, set normally at 1, serves as the inverter when data is entered parallel with the preset inputs. This switch must be set to 0 when entering data serially. The LEDs are connected to the Q output of each flip-flop to indicate the data shifting in the register.

After you have the two 7476 ICs wired-up, you can refer to Figure 5-1 and Table 5-1 for a better understanding of the shift register and apply the following test procedures:

To load register:
1. Set SWE low (1st J input)
2. Set SWG high (1st K input)
3. Momentarily set the desired preset switches A-D to low (normally high)

To shift data:
1. Operate "0" T.P. switch or use clock output set to your desired speed.

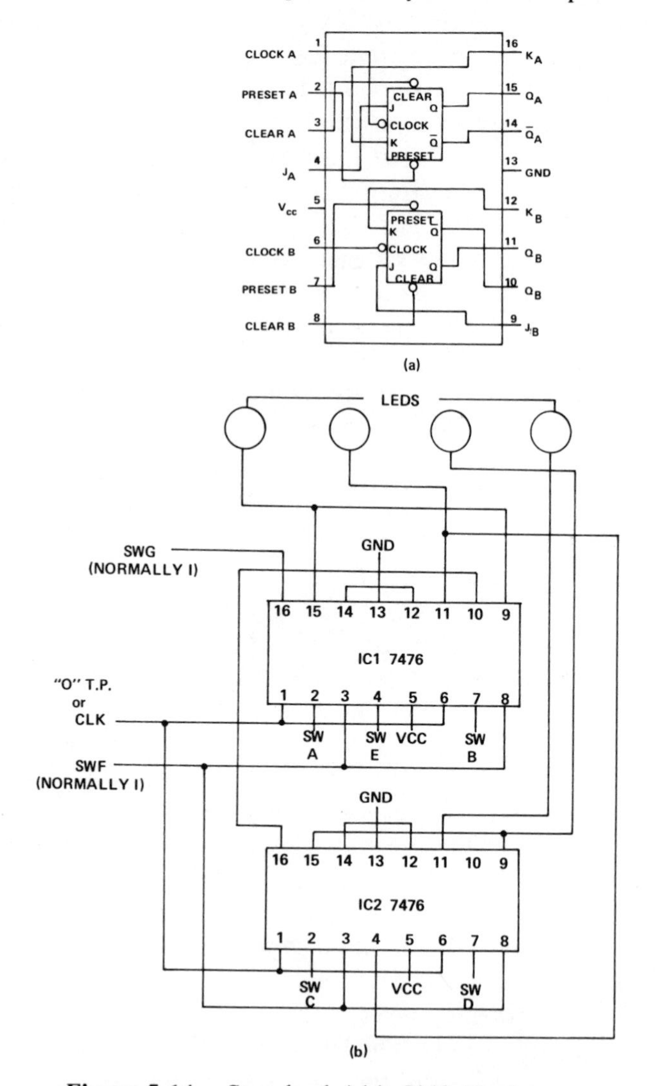

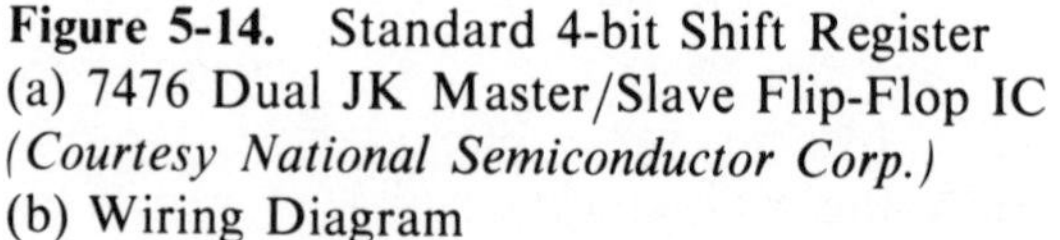

Figure 5-14. Standard 4-bit Shift Register
(a) 7476 Dual JK Master/Slave Flip-Flop IC
(Courtesy National Semiconductor Corp.)
(b) Wiring Diagram

5-6B Ring Counter

Using the standard shift register constructed in Section 5-6A, remove the wire from SWG to pin 16 of IC1 and add the two wires shown in Figure 5-15. Set the clock at the slowest speed and momentarily operate SWA to a low to enable one bit to enter the register. Refer to Figure 5-5 for the ring counter operation.

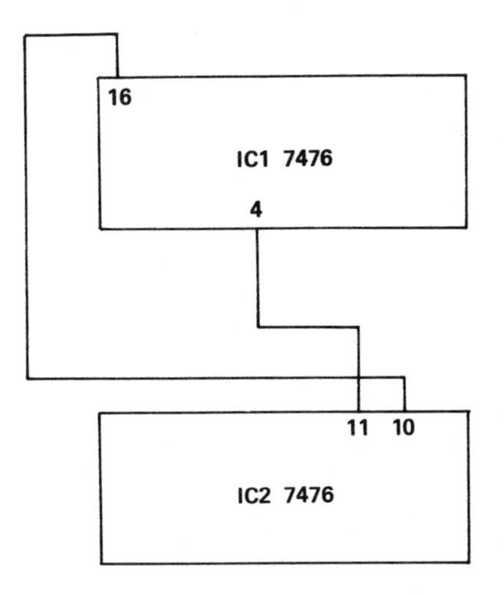

Figure 5-15. Ring Counter Wiring Diagram

5-6C Switch-Tail Ring Counter

Still using the standard shift register, change the two wires added in Section 5-6B to that shown in Figure 5-16. Clear the register by momentarily operating SWF and then the counter should start counting automatically. Refer to Figure 5-6 for the switch-tail ring counter operation. If you want to wire-up the decoder to provide eight pulses, you can use a 7495 4-bit right-shift/left-shift register, a 7404 hex inverter to form the $\overline{Q}$ outputs and two 7408 quad 2-input AND gates ICs. The actual wiring is left to your ingenuity.

5-6D Binary Up-Counter

A 4-bit binary up-counter is fabricated into the 7493 IC as shown in Figure 5-17. Notice that flip-flop A is not wired internally to the other flip-flops. This arrangement provides the 7493 as a divide-by-two counter with the input at pin 14 and a divide-by-eight counter with the input at pin 1. To have a 4-bit counter, pin 12 must be connected to pin 1, with the input pulses applied to pin 14, which produces frequency divisions of 2, 4, 8 and 16 at outputs A, B, C and D respectively. Pins 2 (R_{01}) and 3 (R_{02}) are wired to ground to enable the counter to count up to 16. These inputs will be used more extensively in the next section. You may leave the binary up-counter wired up for the operations in the next section. Refer to Figure 5-8 for the binary up-counter operation.

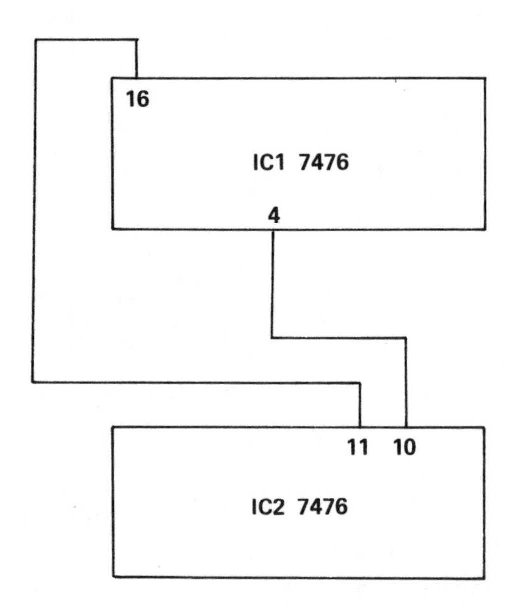

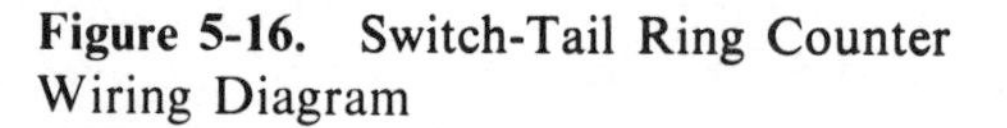
Figure 5-16. Switch-Tail Ring Counter Wiring Diagram

5-6E Modulo Counters

The 7493 IC can be used to construct a modulo counter of from 2 to 16 with the use of inputs R_{01} and R_{02} and in some cases an additional 7408 AND gate IC or 7400 NAND-inverter gate IC as shown in Figure 5-18. Using the counter as wired in the last section, remove the ground from pins 2 and 3 and modify the circuit to your choice of MOD counter with the aid of Table 5-2. Refer to Figure 5-13 for the operation of a MOD counter with reset control.

Type of MOD Counter	Binary Count Before Reset	Binary Rest	Outputs Connected to: $R_{O(1)}$	$R_{O(2)}$	Remarks:
2	01	10	"O"	"O"	Use output A only
3	10	11	A	B	-
4	11	100	C	C	Join R_O (1) and R_O (2)
5	100	101	A	C	-
6	101	110	B	C	-
7	110	111	A	BC	Use AND Gate for B and C
8	111	1000	D	D	Join R_O (1) and R_O (2)
9	1000	1001	A	D	-
10	1001	1010	D	B	-
11	1010	1011	A	BD	Use AND Gate for B and D
12	1011	1100	C	D	-
13	1100	1101	A	CD	Use AND Gate for C and D
14	1101	1110	B	CD	Use AND Gate for C and D
15	1110	1111	AB	CD	Use AND Gate for A and B and C and D
16	1111	0000	"O"	"O"	Normal Reset

Table 5-2 Modulo Counter Connections Using A 7493 4-Bit Binary Counter

7490 Decade Counter

A complete MOD 10 counter is fabricated into the 7490 decade counter IC as shown in Figure 5-19. This IC has the same feature as the 7493 IC with flip-flop A not being internally connected. The BD

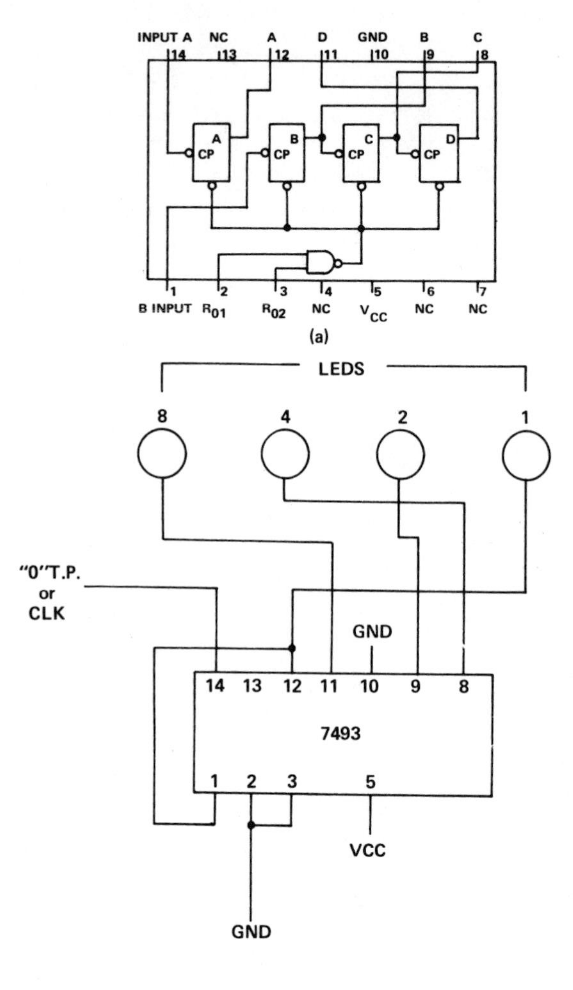

Figure 5-17. 4-bit Binary Up Counter (a) 7493 4-bit Binary Counter *(Courtesy National Semiconductor Corp.)* (b) Wiring Diagram

input produces a divide-by-five operation, while input A allows a divide-by-two, if desired. When output A is connected to input BD, the counter functions in the normal MOD 10 mode. Inputs R_{01} and R_{02} also control the clearing of the counter, while inputs R_{g1} and R_{g2} control the counter by setting it to 9 for special applications.

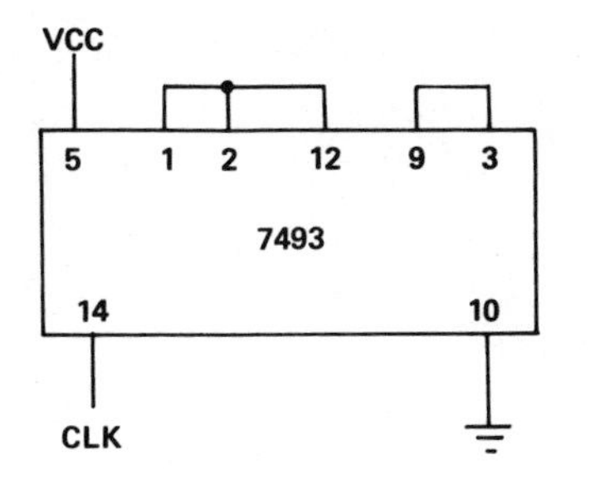

(a)

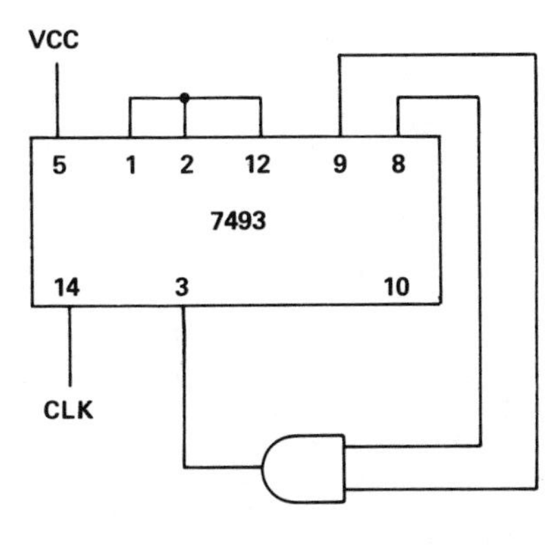

(b)

Figure 5-18. Modulo Counter Construction
(a) MOD-3 Counter (b) MOD-7 Counter

7492 Divide-by-12 Counter

The 7492 divide-by-12 counter is similar to the 7493 4-bit binary counter, except that with internal feedback it becomes a MOD 12 counter. When output A is connected to the BC input, the counter operates in the normal mode. When the input frequency is applied to the BC input, a frequency division of 3 and 6 results on the C and D outputs respectively. Inputs R_{01} and R_{02} are the clear control as shown in Figure 5-20.

5-6F Binary Up/Down Counter

The 74193 up/down binary counter is a 4-bit counter that will count up from 0-15 or count down from 15-0 depending on which input is used, i.e., down count, pin 4 or up count, pin 5, as shown in Figure 5-21a. The parallel load feature allows this counter also to be used as a modulo counter if it is preset with the proper number. There are two outputs, borrow and carry, which are connected to the clock inputs of subsequent counters for counting numbers greater than 15.

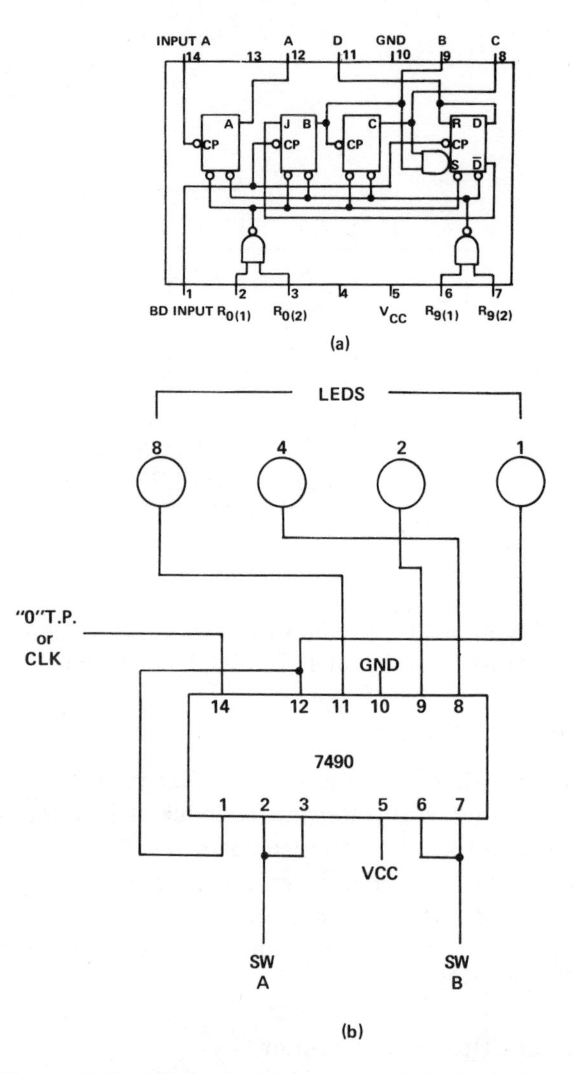

Figure 5-19. Decade Counter (a) 7490 Decade Counter *(Courtesy National Semiconductor Corp.)* (b) Wiring Diagram

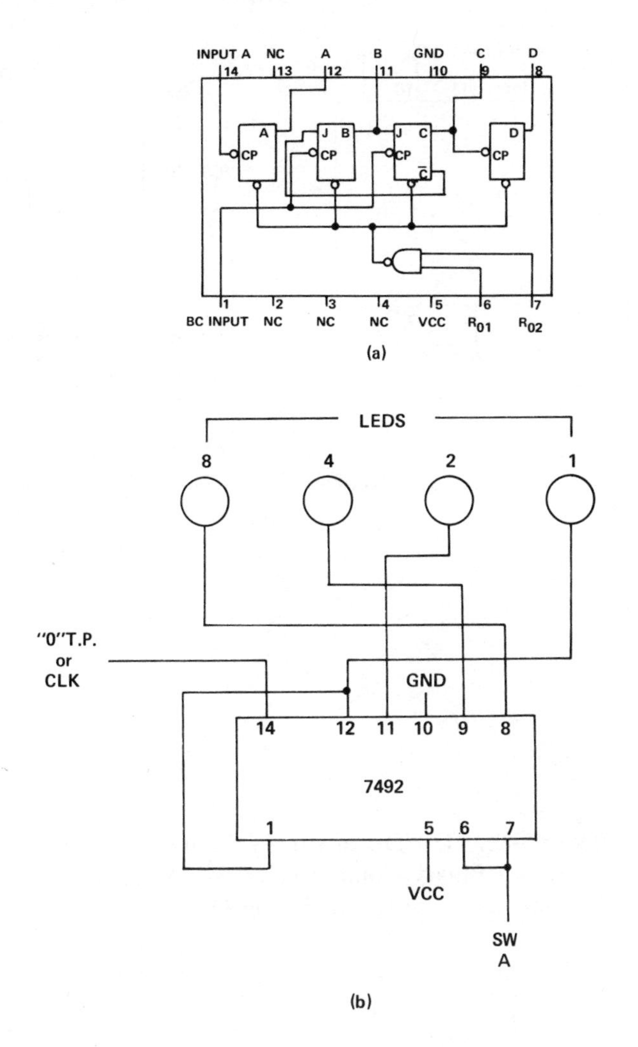

Figure 5-20. Divide-by-12 Counter (a) 7492 Divide-by-12 Counter *(Courtesy National Semiconductor Corp.)* (b) Wiring Diagram

In the wiring diagram shown in Figure 5-21b, switches A-D are set high for the desired number to be entered into the register. Switch E, which is normally high, is momentarily set low in order to load the register. Now, depending on which input, pin 4 or pin 5, is "0" pulsed the counter will count down or count up respectively. Pin 14 requires a 1 pulse to clear the register. Refer to Figure 5-11 for the logic diagram of the 74193 IC.

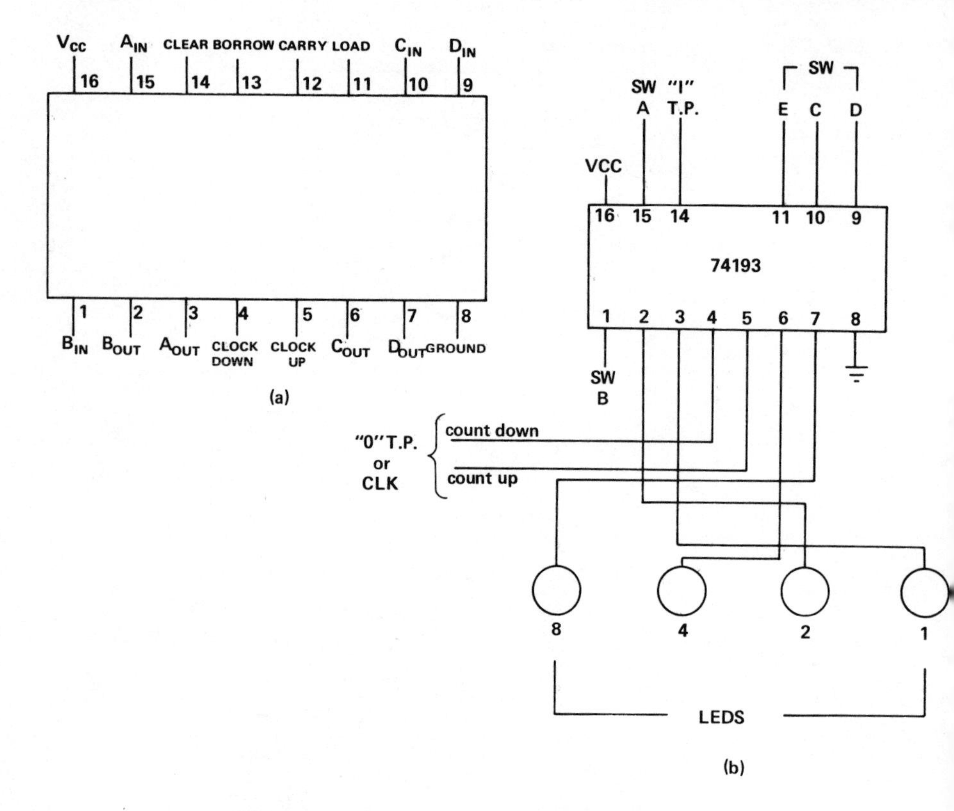

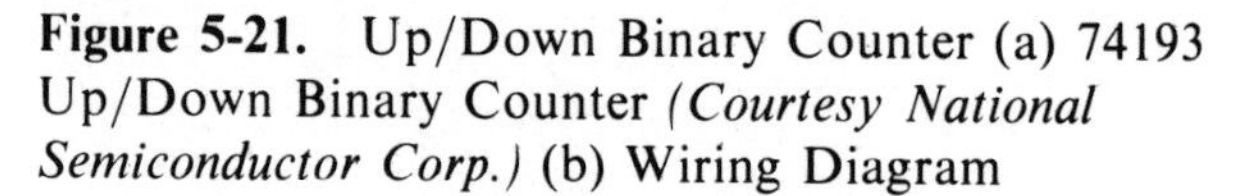
Figure 5-21. Up/Down Binary Counter (a) 74193 Up/Down Binary Counter *(Courtesy National Semiconductor Corp.)* (b) Wiring Diagram

CHAPTER 6

Adding and Subtracting with Digital Circuits

The binary adder is the most versatile digital circuit for performing arithmetic operations. The adder is capable of addition, subtraction, multiplication and division.

A binary subtractor can be constructed from basic logic gates. However, an adder using complementing techniques, as described in Chapter 1, can very easily perform subtraction and, therefore, eliminates additional circuitry in a digital system employing arithmetic operations. Multiplication is accomplished with an adder by a method of successive adding and shifting, similar to the way in which you multiply with pencil and paper. In a like manner, the adder can perform division by comparing numbers and successive subtracting and shifting also as you do with pencil and paper.

In this chapter, you will learn how the half adder is combined into a full adder used for addition and subtraction and how comparator circuits indicate the relative magnitude of two numbers.

In Project # 6, you will be able to construct a serial adder, a parallel adder used for adding and subtracting, and test a 4-bit magnitude comparator.

6-1 ADDITION WITH A BINARY ADDER

Half Adder

Perhaps, the best way to understand how a binary adder works is first to study the half adder. A half adder is capable of adding only two 1-bit binary numbers. The logic symbol for a half adder is shown in Figure 6-1.

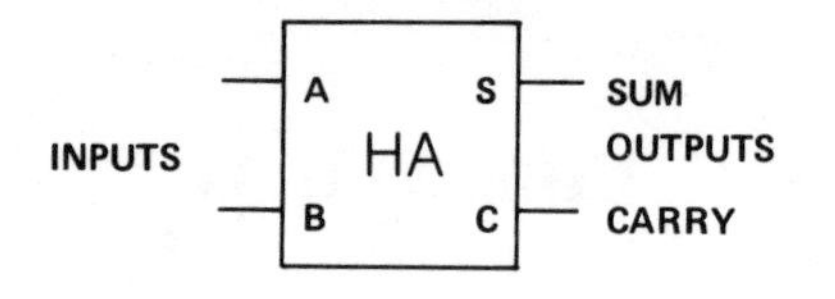

(a)

INPUTS		OUTPUTS	
A	B	S	C
0	0	0	0
0	1	1	0
1	0	1	0
1	1	0	1

(b)

$$S = A\bar{B} + \bar{A}B$$
$$C = AB$$

(c)

Figure 6-1. Binary Half Adder
(a) Logic Symbol (b) Truth Table (c) Formulas

Inputs A and B represent the two binary digits to be added. The outputs S(sum) and C(carry) are the two resulting conditions from a binary addition and are shown in the truth table. You may want to return to Section 1-4 and review binary addition at this time. The formula for a half adder gives a clue to its logic gate implementation. The simplest logic diagram could have the sum output be an exclusive-OR gate, while the carry output represents an AND gate as shown in Figure 6-2a.

The rest of the figure shows other logic gate configurations. It would be helpful to get a piece of scratch paper, redraw each diagram, apply 1's and 0's to the inputs and with your knowledge of logic gates, prove the truth table.

Full Adder

The half adder is very limited as it has no provision for entering a carry back into the adder for more than a single two-bit column of numbers. Combining two half adders with an OR function creates a full adder capable of not only adding the two standard A and B inputs, but it can also add the carry that results from a previous column of addition. The full adder has three inputs, i.e., A, B and Ci (carry in) and two outputs So (sum out) and Co (carry out) as shown in Figure 6-3.

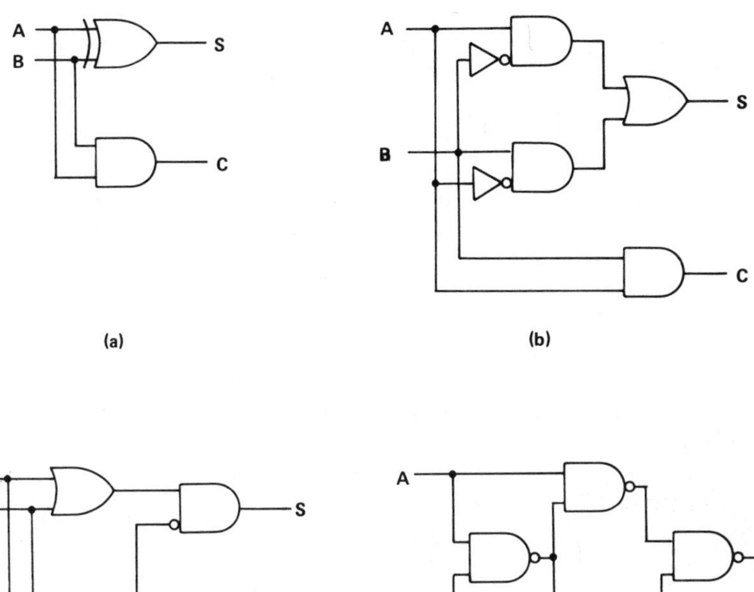

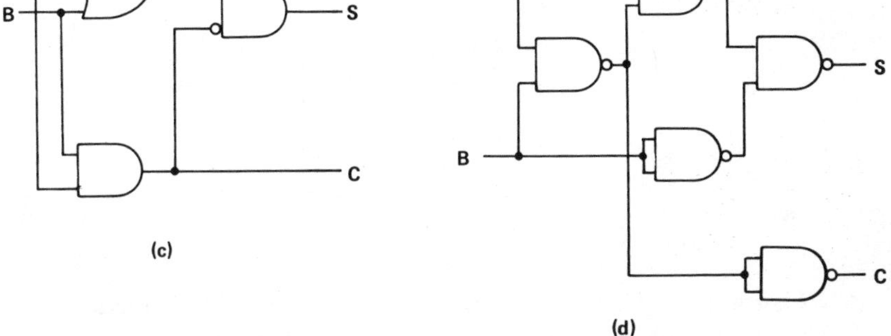

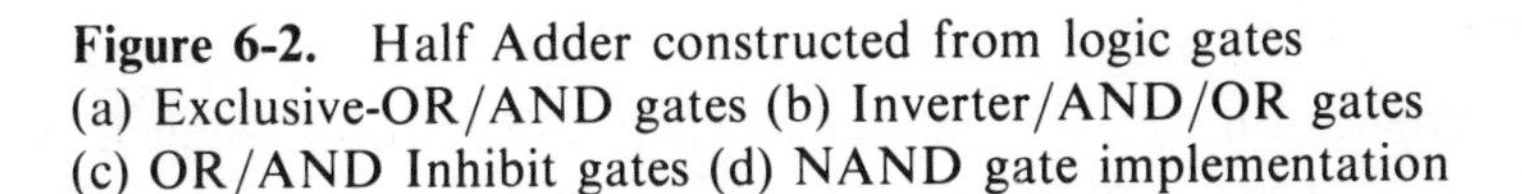

Figure 6-2. Half Adder constructed from logic gates
(a) Exclusive-OR/AND gates (b) Inverter/AND/OR gates
(c) OR/AND Inhibit gates (d) NAND gate implementation

You may want to draw the logic gate diagrams for each half adder with the combined OR gate and then prove the truth table for the full adder. Essentially, the truth table and formulas state that the So output will be a 1 if only one input is high, the Co output will be a 1 if any two inputs are high, and both So and Co outputs will be 1, if all inputs are high.

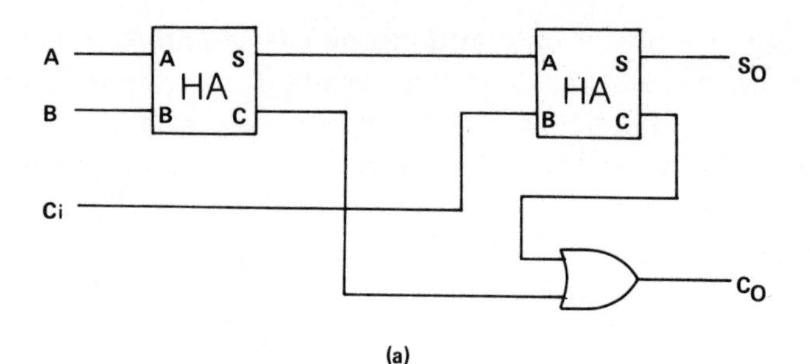

(a)

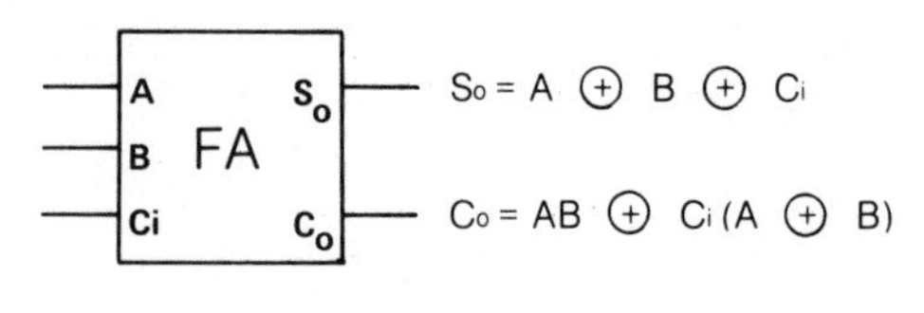

(b) (c)

INPUTS			OUTPUTS	
A	B	Ci	So	Co
0	0	0	0	0
0	0	1	1	0
0	1	0	1	0
0	1	1	0	1
1	0	0	1	0
1	0	1	0	1
1	1	0	0	1
1	1	1	1	1

(d)

Figure 6-3. Full Adder (a) Half Adders and OR Function (b) Logic Symbol (c) Formulas (d) Truth Table

Serial Addition

Operation of a full adder is easily understood when it is connected as a serial adder shown in Figure 6-4.

The addition process is similar to the manual operation using pencil and paper. Each column, beginning with the LSB, is added in a step-by-step fashion. Registers A and B are shifted into the adder bit-by-bit and the sum is shifted into Register C. Any carry that results is temporarily stored in the delay flip-flop and then added with the next column of numbers that is shifted into the adder. The add control circuit isn't necessary, but aids in showing the shifting and adding operation of the register table in Figure 6-4b.

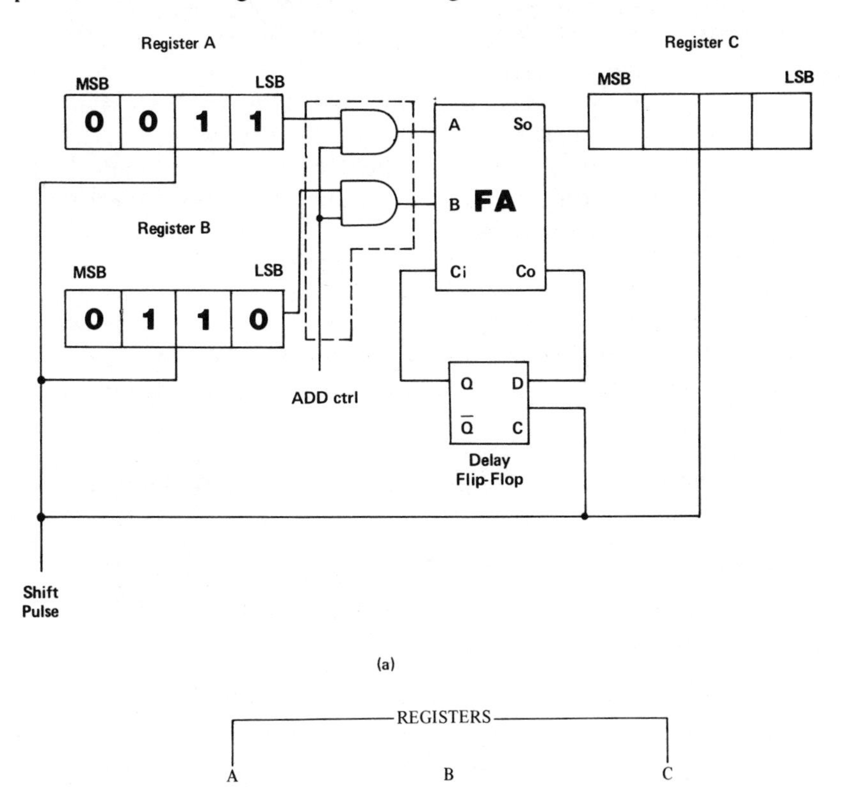

(a)

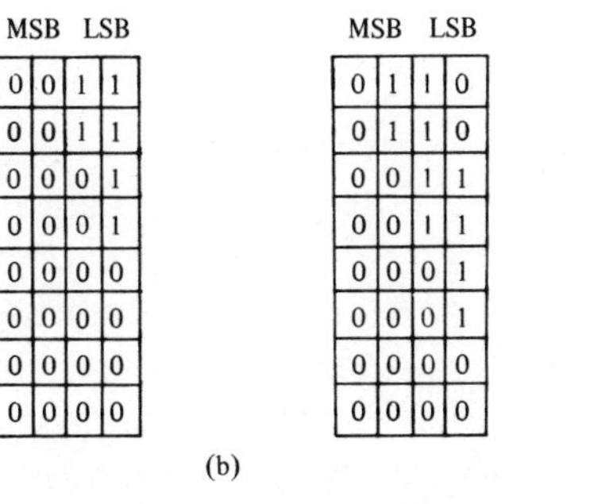
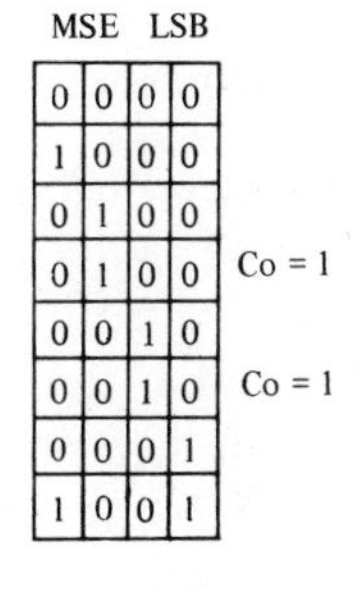

	REGISTERS A (MSB LSB)	REGISTERS B (MSB LSB)	REGISTERS C (MSE LSB)	
LOAD	0 0 1 1	0 1 1 0	0 0 0 0	
ADD	0 0 1 1	0 1 1 0	1 0 0 0	
SHR #1	0 0 0 1	0 0 1 1	0 1 0 0	
ADD	0 0 0 1	0 0 1 1	0 1 0 0	Co = 1
SHR #2	0 0 0 0	0 0 0 1	0 0 1 0	
ADD	0 0 0 0	0 0 0 1	0 0 1 0	Co = 1
SHR #3	0 0 0 0	0 0 0 0	0 0 0 1	
ADD	0 0 0 0	0 0 0 0	1 0 0 1	

(b)

Figure 6-4. Serial Addition
(a) Logic Diagram (b) Register Table

The adding operation is as follows:

Binary number 3 is loaded into Register A.
Binary number 6 is loaded into Register B.

Add ctrl line goes high and the LSB of registers A & B and input Ci are added (1+0+0=So=1 Co=0). The So output places a 1 in the MSB of Register C.
SHR #1 causes the data in Registers A, B and C to shift right one place.
Add ctrl line goes high and the LSB of registers A, B and input Ci are added (1+1+0=So=0 CO=1). The Co output places a 1 on the D input of the delay flip-flop which will termporarily store the carry until the next place or column is added.
SHR #2 causes the data in Registers A, B and C to shift right one place again and sets the delay flip-flop.
Add ctrl line goes high and the LSB of registers A, B and input Ci are added (0+1+1=So=0 Co=1). The Co output again places a 1 at the input of the delay flip-flop.
SHR #3 causes the data in Registers A, B and C to shift right one more place and sets the delay flip-flop again.
Add ctrl line goes high and the LSB of registers A and B and input Ci are added for the last time (0+0+1=So=1 Co=0). The answer 1001 is now in register C.

Notice that for this particular circuit you need four Add ctrl pulses and three SHR pulses to perform a complete 4-bit binary addition operation.

Parallel Addition

Serial addition requires more time to shift the bits through the adder than using parallel addition shown in Figure 6-5.

Parallel addition requires a full adder for each bit of the binary number. The LSB of Registers A and B are fed into FA0. Likewise, each ascending position of Registers A and B is fed into separate full adders. The Co outputs are fed to the next highest order Ci inputs with the Co output of FA3 being used as an overflow indicator where a number larger than binary 1111 may result from an add operation. The So outputs of the adders are fed to the corresponding positions of Register C.

Register C can be eliminated in both serial and parallel addition circuits. The So outputs are fed back to Register A and this register is commonly referred to as the accumulator. You will be able to wire-up this type of circuit given in Project #6.

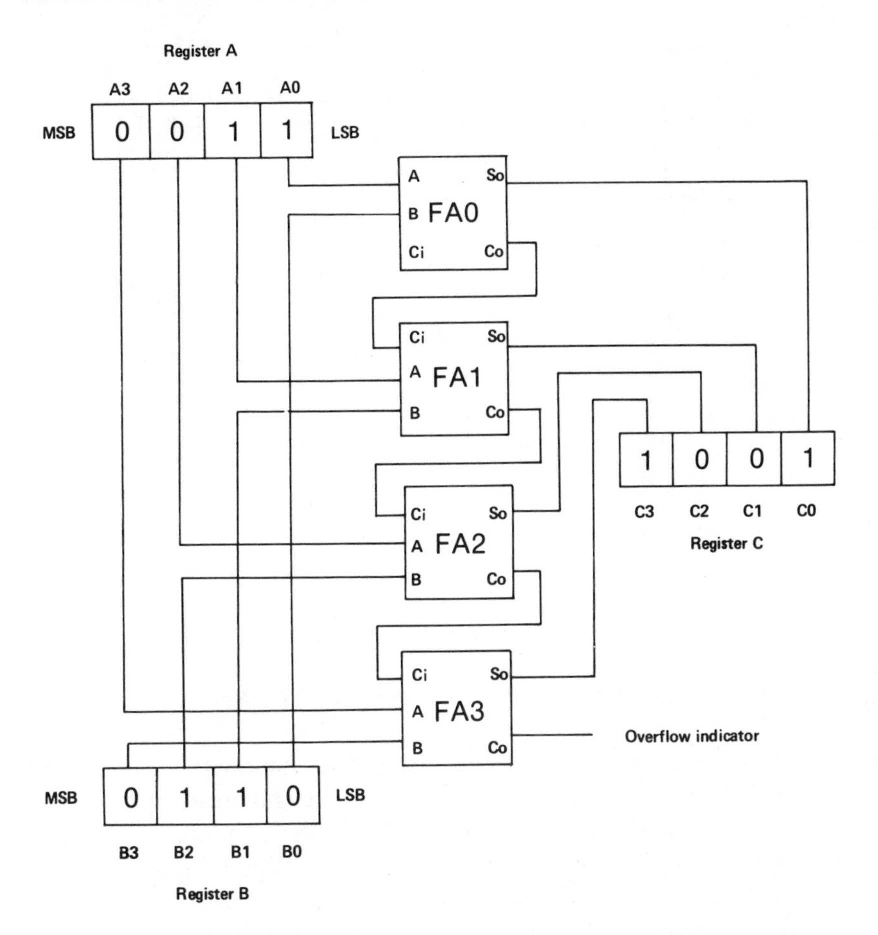

Figure 6-5. Parallel Addition

6-2 SUBTRACTION BY COMPLEMENTING AND ADDING

In Section 1-4 you learned how to subtract binary numbers by complementing and adding. There are various ways of complementing a number with digital circuits, one of which is shown in Figure 6-6.

This circuit is a serial adder that uses the 2's complement for subtraction. When the mode ctrl is low the exclusive-OR gate allows the contents of Register B to be shifted into the adder unaffected. However, when the mode ctrl is high, the contents of Register B are complemented as they shift into the adder. Also, when the mode ctrl goes high, the one-shot multivibrator sets the delay flip-flop which puts a 1 on the Ci input to accomplish the 2-s complement operation.

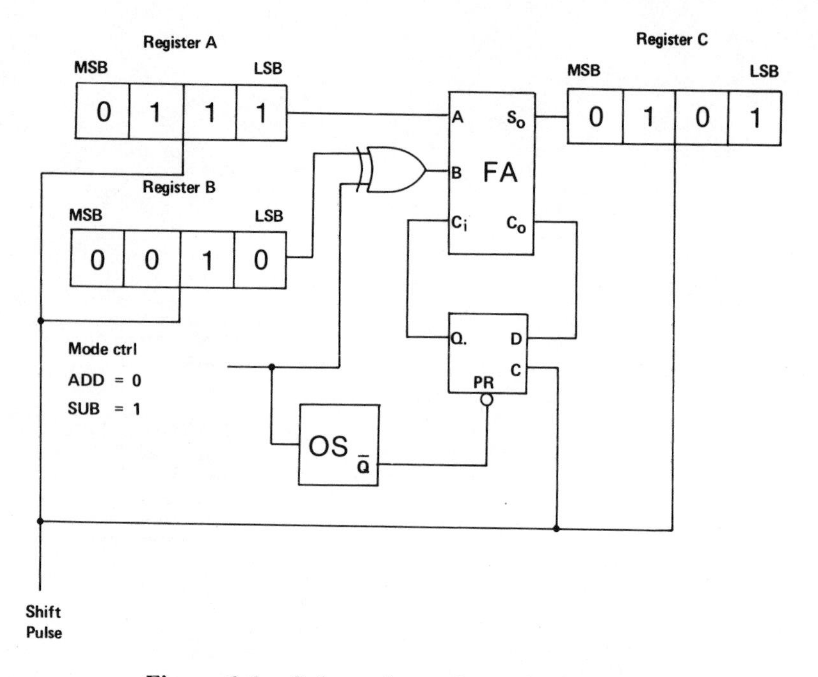

Figure 6-6. Subtraction using a Serial Adder

Other basic circuits that can be used to complement a number when subtraction is required of an adder are shown in Figure 6-7.

When the add line of AND gate 1 is high in Figure 6-7a, the Q output of the register is shifted into the adder. When the SUB line of AND gate 2 is high the $\overline{Q}$ output of the register is shifted into the adder. The add and sub should never be high at the same time. This same arrangement can be used for each bit of the binary word for parallel addition/subtraction as shown in Figure 6-7b.

6-3 COMPARATOR USED FOR COMPARING THE MAGNITUDE OF TWO NUMBERS

A simple one-bit comparator is shown in Figure 6-8. An output of 1 is obtained when inputs A and B are the same. When the inputs are not the same, a 0 appears at the output.

Using four of these simple comparators and additional decode logic networks, a 4-bit magnitude comparator could be fabricated into one IC as shown in Figure 6-9.

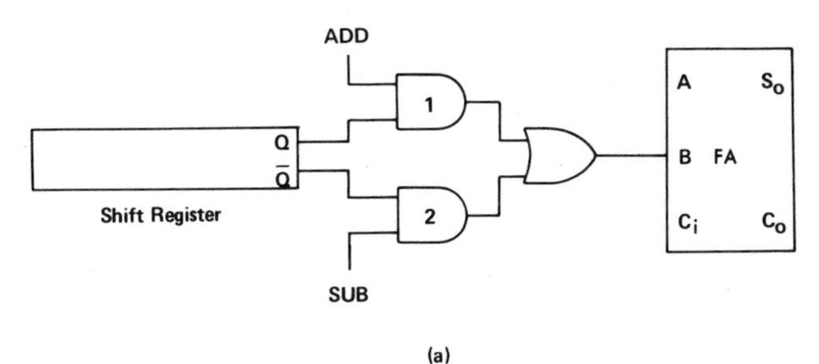

(a)

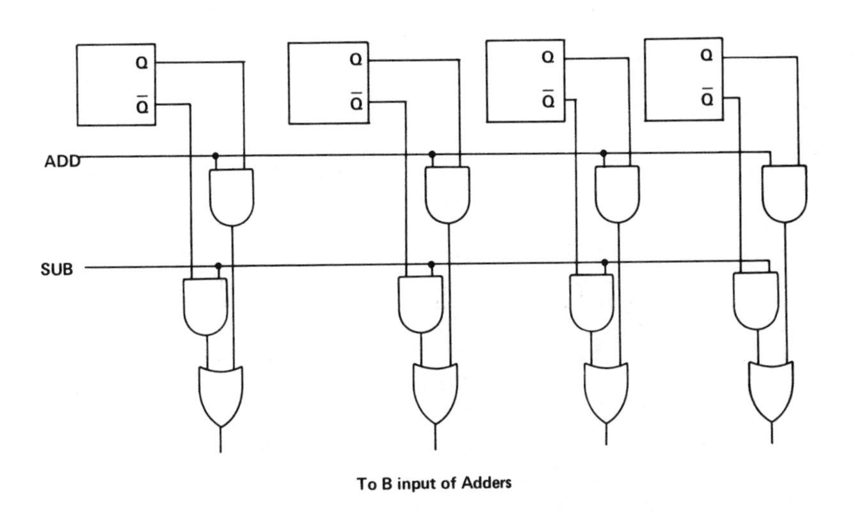

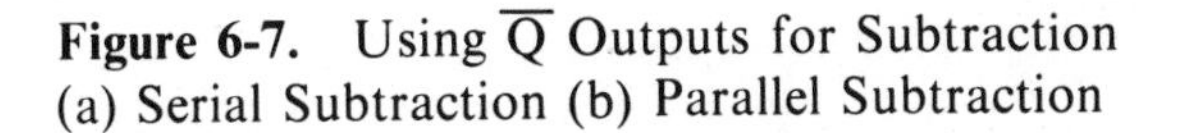
Figure 6-7. Using $\overline{Q}$ Outputs for Subtraction
(a) Serial Subtraction (b) Parallel Subtraction

The 7485 4-bit magnitude comparator has four inputs for Register A, four inputs for Register B, and three outputs that indicate whether A is larger than B, A is less than B, or A is equal to B. The three cascading inputs (A<B, A = B, A>B) are used with other 7485 ICs where larger words are being compared. If only 4-bit words are being compared, the A=B input should be wired to a high while the A<B and A>B inputs should be wired to a low.

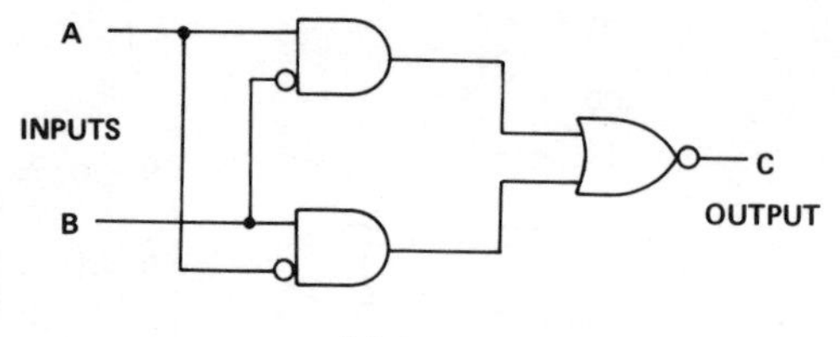

(a)

A	B	C
0	0	1
0	1	0
1	0	0
1	1	1

(b)

$$\left.\begin{array}{l} AB = C \\ \overline{A}\overline{B} = C \end{array}\right\} C = 1 \text{ WHEN } A = B$$

(c)

Figure 6-8. Simple Comparator (a) Logic Diagram (b) Truth Table (c) Formula

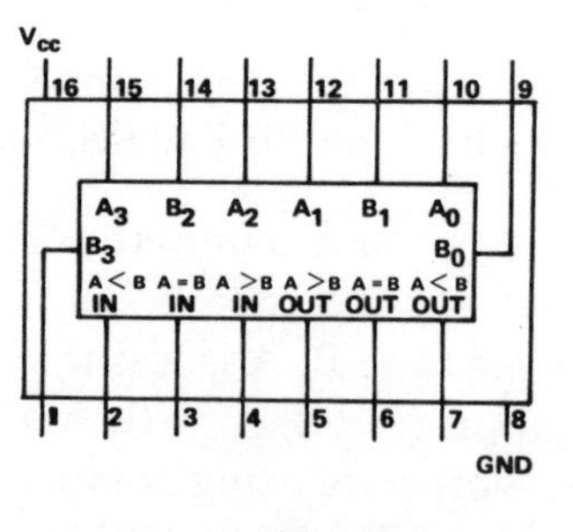

(a)

Figure 6-9. 7485 4-bit Magnitude Comparator (a) Connection Diagram (b) Truth Table *(Courtesy National Semiconductor Corp.)*

COMPARING INPUTS				CASCADING INPUTS			OUTPUTS		
A3, B3	A2, B2	A1, B1	A0, B0	A > B	A < B	A = B	A > B	A < B	A = B
A3 > B3	X	X	X	X	X	X	H	L	L
A3 < B3	X	X	X	X	X	X	L	H	L
A3 = B3	A2 > B2	X	X	X	X	X	H	L	L
A3 = B3	A2 < B2	X	X	X	X	X	L	H	L
A3 = B3	A2 = B2	A1 > B1	X	X	X	X	H	L	L
A3 = B3	A2 = B2	A1 < B1	X	X	X	X	L	H	L
A3 = B3	A2 = B2	A1 = B1	A0 > B0	X	X	X	H	L	L
A3 = B3	A2 = B2	A1 = B1	A0 < B0	X	X	X	L	H	L
A3 = B3	A2 = B2	A1 = B1	A0 = B0	H	L	L	H	L	L
A3 = B3	A2 = B2	A1 = B1	A0 = B0	L	H	L	L	H	L
A3 = B3	A2 = B2	A1 = B1	A0 = B0	L	L	H	L	L	H

NOTE: H = high level, L = low level, X = irrelevant.

(b)

Figure 6-9. *(Continued)*

SUMMARY

Binary adders can be built from basic logic gates or can be found in IC form. The full adder has three inputs and two outputs and its operation can be summarized thus:

with only one input high —outputs So=1 and Co=0
with any two inputs high —outputs So=0 and Co=1
with all inputs high —outputs So=1 and Co=1

By complementing one register and setting the carry input, the full adder can be used for subtraction with 2's complement.

The magnitude comparator can sample the contents of two registers and give an indication of which one is greater or if they both are equal. This information can be used in other digital circuits to perform certain operations.

By now you can see the time saved in building or testing ICs that are more complex, rather than complex circuits made from basic logic gates and single flip-flops.

6-4 PROJECT #6: CONSTRUCTING AND TESTING DIGITAL ARITHMETIC CIRCUITS

In Project #6, you will begin to work with more complex integrated circuits. The impact of how these more complex ICs save time in construction and testing will be realized if you first assemble the half adders shown in Figure 6-2 into full adders using basic logic gate ICs. For example, a full adder made up of two half adders shown in Figure 6-2a will require two ICs, a 7486 quad EXCLUSIVE-OR gate IC and a 7408 quad 2-input AND gate IC. This combination results in a single-bit adder capable only of serial addition when multiple-bit words are used.

In contrast, the 7483 4-bit full adder IC used in the project can be wired to produce two single-bit full adders used for serial addition or a 4-bit full adder for parallel addition.

You will have to refer to Project # 5 for detailed instructions on how the registers are wired for the arithmetic operations, but remember to use the same precautions in handling and wiring ICs that were given in previous projects.

6-4A Serial Addition

The 7483 is a 4-bit binary full adder usually used in parallel addition, but which can also be used as a dual single-bit binary adder as shown in Figure 6-10.

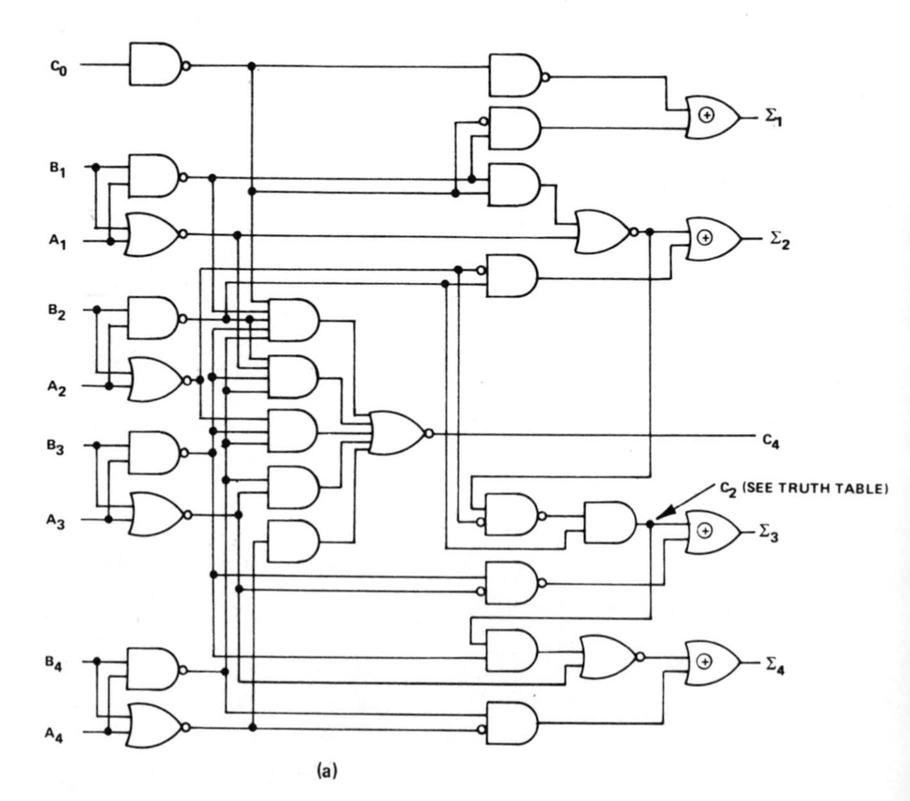

(a)

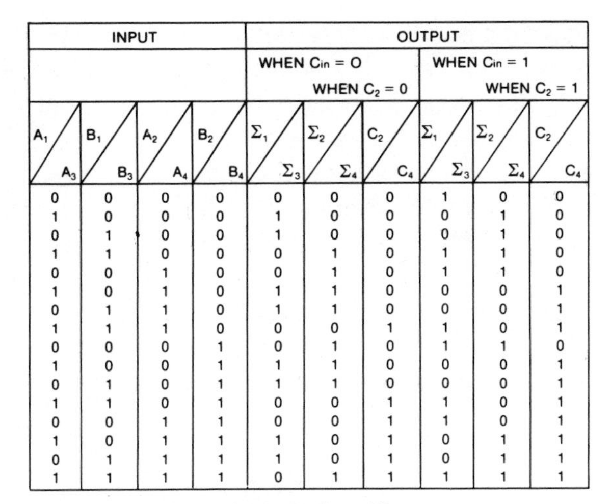

INPUT				OUTPUT					
				WHEN $C_{in} = 0$ / WHEN $C_2 = 0$			WHEN $C_{in} = 1$ / WHEN $C_2 = 1$		
A_1 / A_3	B_1 / B_3	A_2 / A_4	B_2 / B_4	Σ_1 / Σ_3	Σ_2 / Σ_4	C_2 / C_4	Σ_1 / Σ_3	Σ_2 / Σ_4	C_2 / C_4
0	0	0	0	0	0	0	1	0	0
1	0	0	0	1	0	0	0	1	0
0	1	0	0	1	0	0	0	1	0
1	1	0	0	0	1	0	1	1	0
0	0	1	0	0	1	0	1	1	0
1	0	1	0	1	1	0	0	0	1
0	1	1	0	1	1	0	0	0	1
1	1	1	0	0	0	1	1	0	1
0	0	0	1	0	1	0	1	1	0
1	0	0	1	1	1	0	0	0	1
0	1	0	1	1	1	0	0	0	1
1	1	0	1	0	0	1	1	0	1
0	0	1	1	0	0	1	1	0	1
1	0	1	1	1	0	1	0	1	1
0	1	1	1	1	0	1	0	1	1
1	1	1	1	0	1	1	1	1	1

Note 1: Input conditions at A_1, A_2, B_1, B_2, and C_{in} are used to determine outputs Σ_1 and Σ_2, and the value of the internal carry C_2. The values at C_2, A_3, B_3, A_4, and B_4, are then used to determine outputs Σ_3, Σ_4, and C_4.

(b)

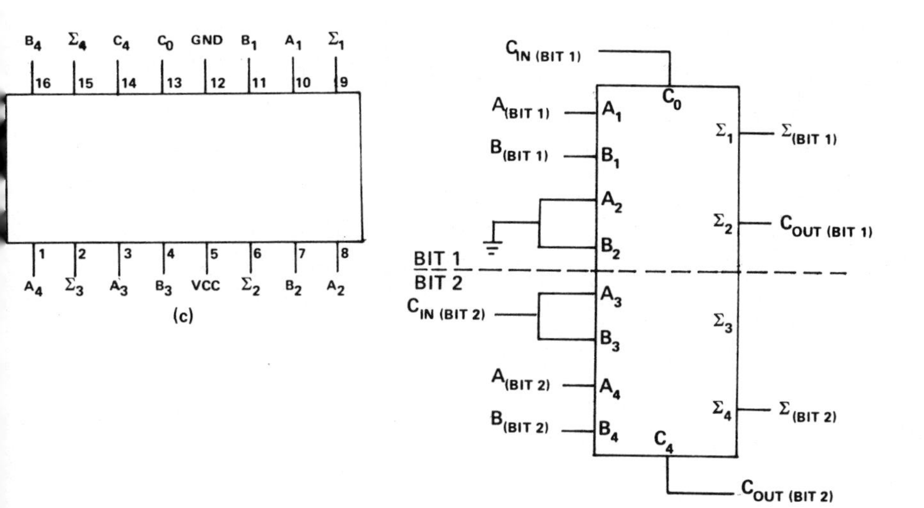

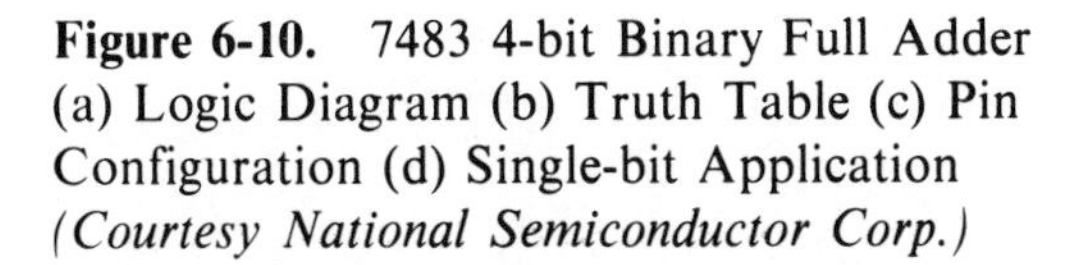

Figure 6-10. 7483 4-bit Binary Full Adder (a) Logic Diagram (b) Truth Table (c) Pin Configuration (d) Single-bit Application *(Courtesy National Semiconductor Corp.)*

When used in a parallel operation, one 4-bit register is connected to the A inputs which are weighted A1=1, A2=2, A3=4, A4=8, and the other 4-bit register is connected to the B inputs which are weighted in similar fashion. There is a carry input at pin 13 labeled C_0 and a carry output at pin 14 labeled C_4. When the 7483 is used only with 4-bit numbers, the C_0 input should be grounded. For numbers larger than 4-bits the C_0 and C_4 connections are used to connect the adders together similar to that shown in Figure 6-5.

The 7483 is implemented to serve as a dual single-bit full adder as shown in Figure 6-10c. The upper half will be used to test the serial addition operation of Figure 6-11. Two 7495 shift registers are fed

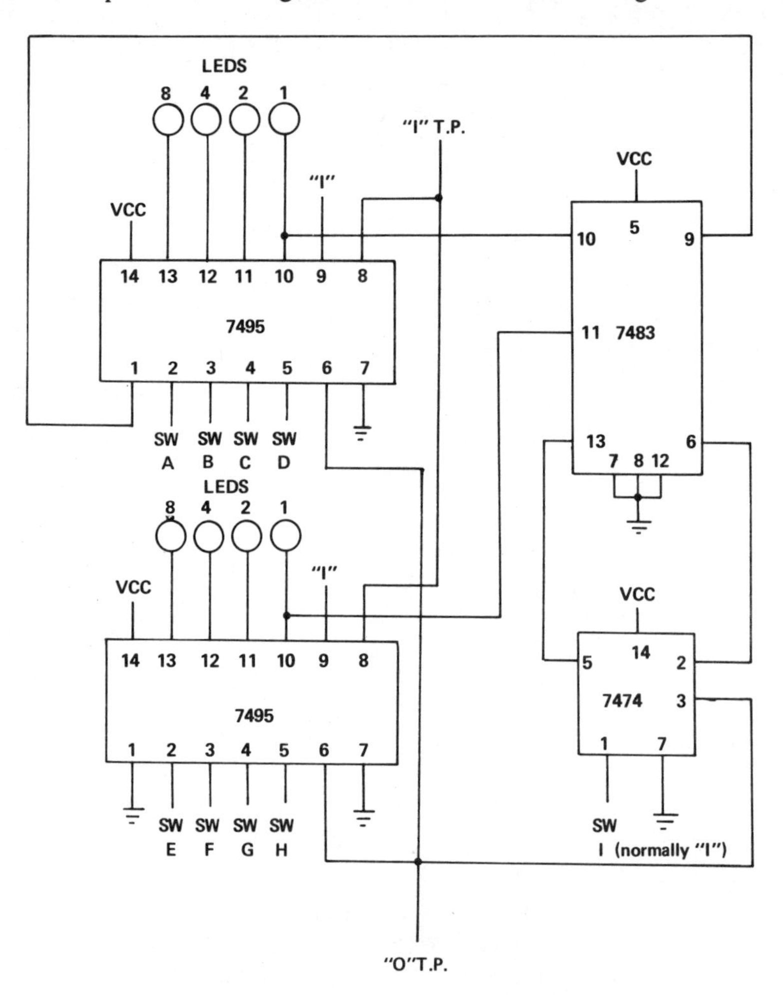

Figure 6-11. Serial Addition Wiring Diagram

into the appropriate inputs of the 7483 full adder and one half of a 7474 dual D flip-flop is used as the carry delay.

The output of the adder is fed back into register A via the serial input of the 7495 IC. As the two numbers are shifted into the adder to be added, the sum is shifted back into Register A which serves as the accumulator. Figure 5-3b shows the logic diagram of the 7495 IC.

The 7495 registers normally shift right when the mode control is low and the clock 1 R-shift input is "1" pulsed, and shift left when the mode control is high and the clock 2 L-shift input is "1" pulsed. Because of the lack of more switches on the breadboard/tester, the clock 1 input is tied high and data is entered (shift left) when a 1 is placed on clock 2. The shift right operation is accomplished when the mode control input goes low. Refer to Figure 6-4 for a better understanding of serial addition and apply the following test procedures:

To Load Registers:

1. Set desired input switches A-H for each register respectively.
2. Operate "1" T. P. switch
3. Return input switches A-H

For Add Operation:

1. Operate SWI through one cycle from normally high to low and back to high. (This clears the delay flip-flop.)
2. Operate the "0" T. P. switch four times. (Sum should be in register A)

To Clear Accumulator:

1. Operate "1" T. P. switch with switches A-H in low position.

6-4B Parallel Addition and Subtraction

By rewiring the 7483 full adder for the normal operation and including a 7486 Quad exclusive-OR gate, you are able to add and subtract, using the same circuit as shown in Figure 6-12. The LEDs are removed from the registers and only the output of the adder is monitored. Switch I acts as the mode control for addition and subtraction. When SWI is low, the adder accomplishes normal addition. When SWI is high, the bottom register is complemented via the 7486 Exclusive-OR gates, C is set and subtraction is accomplished. Once the registers are loaded, the adder produces an output as indicated by the LEDs. For a better understanding, refer to Figure 6-5, 6-6 and 6-7 and apply the following test procedures:

To load registers and Add:

1. Set SWI low
2. Set desired input switches A-H for each register respectively.
3. Operate "1" T.P. switch (sum indicated by LEDs)

To clear registers and adder:

1. Set input switches A-H in low position
2. Operate "1" T.P. switch

To load registers and subtract:

1. Set SWI high
2. Set desired input switches A-H for each register respectively.
3. Operate "1" T.P. switch (difference indicated by LEDs)

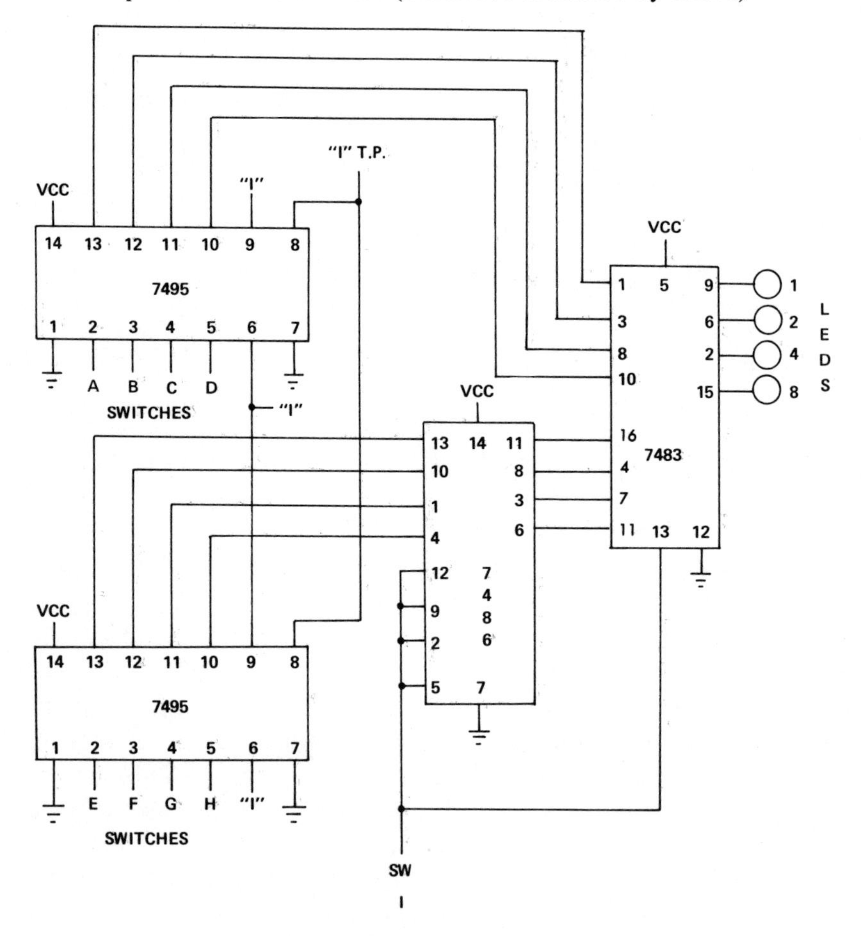

Figure 6-12. Parallel Addition and Subtraction Wiring Diagram

6-4C 4-bit Magnitude Comparator

The input switches can be used alone to test the 7485 4-bit magnitude comparator, as shown in Figure 6-13. Input switches A-D are connected to inputs A_0-A_3 respectively to form Register A. Input switches E-H are connected to inputs B_0-B_3 respectively, to form Register B. The output is indicated by the LEDs directly as the contents of the registers change. Refer to Figure 6-9 and test various combinations of numbers to prove A>B, A=B and A<B.

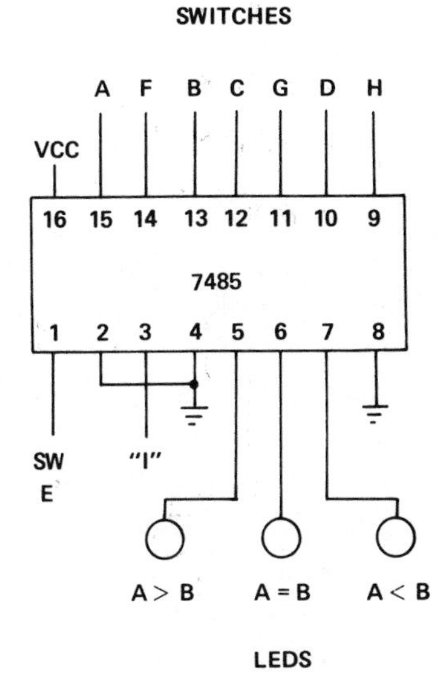

Figure 6-13. 4-bit Magnitude Comparator Wiring Diagram

CHAPTER 7

How Memory Circuits Store Instructions and Data

A memory circuit in digital electronics is a circuit that can store 1 or 0 bit. As seen in Chapter 4, the flip-flop is a basic memory device. A group of flip-flops combined into a register is able to store a group of bits or a complete word. The storage flip-flops in a memory are referred to as memory cells and, depending on how the memory is organized, is able to store many words. Additional circuits such as drivers, decoders, sense amplifiers and control gates are needed so that the proper memory locations can be accessed where data can be stored and retrieved.

There are, of course, many other types of memory devices used with digital circuits, especially in the area of computers. Some of these memory devices include:

Magnetic core, where the memory cell is a tiny donut-shaped magnet.

Magnetic drum, magnetic disk and magnetic tape, where the data is stored as tiny areas of magnetization. These devices must be moving or rotating in order to store and retrieve the data.

Punched card or paper tape, where a hole represents a 1 bit and the absence of a hole represents a 0 bit.

Other memory devices being researched and developed, such as magnetic bubble, laser memories and card capacitor and other, not so widespread devices, such as magnetic rod memory and thin film memory.

Memories in IC form are semiconductor memories which you will study in this chapter. Initially, semiconductor memories were physically larger and more costly than magnetic core memories. However, with the advanced techniques of IC fabrication, some semiconductor memories contain many thousands of bits including the circuitry needed to store and retrieve the data. The smaller semiconductor memories that were developed first, often referred to as scratch pad memories, are useful for storing data temporarily.

Project #7 will provide you with experience in wiring and testing an actual 64-bit scratch pad memory.

7-1 UNDERSTANDING THE RANDOM ACCESS MEMORY (RAM)

The random access memory can select any memory cell or group of cells instantly in its array with the proper addressing inputs. A simplified drawing of a 16-bit memory array, for illustrative purposes, is shown in Figure 7-1.

Each memory cell has an X and Y address line connected to it. By selecting one X line and one Y line, the desired memory cell is accessed and conditioned either to write-in (store) a bit or read-out its contents. For instance, if line X2 and line Y3 go high, memory cell 7 is selected. If we only want to read the contents of memory cell 7, its output of either 0 or 1 will appear at the data output. We can write-in a 0 or 1 into memory cell 7 with the proper control at the data input. With this method of coincident-selection, only 8 lines are needed to select any one of the 16 memory cells. This 16-bit memory array is capable of writing-in and reading-out only 1 bit at a time.

The 7489 64-bit random access read/write memory IC shown in Figure 7-2 is organized as a 16- × 4-bit word RAM capable of storing 4 bits at a time.

Using the linear-select method of addressing, any 4-bit word location in the memory array is selected by decoders when the desired 4-bit address number is placed at the address inputs A0-A3. (Only 3 words of the 16-word array are shown in the figure.) After addressing, data may be either written into or read from the memory. In order to write into the memory both the Memory Enable (ME) and the Write Enable (WE) inputs must be low Data now applied to the four data inputs D_1-D_4will be stored at the addressed location. To read data from a specific location, the ME input must be 0 and the WE input a 1. The data stored at the addressed location will appear at the four sense outputs S_1-S_4 as the complement of the data that was written into the memory. This complemented output is advantageous in some applications as you will see in Chapter 8. Inverters are used at the sense outputs if it is desired to have the data in the original form that was stored into the memory. The 7489 IC has open collector outputs at the sense outputs to permit "wired-OR" capability, which means that a 1000-ohm pull-up resistor is required between Vcc and each S_1-S_4 outputs. This resistor network and the pin rearrangement of the 7489 for easier reading are shown in Figure 7-3.

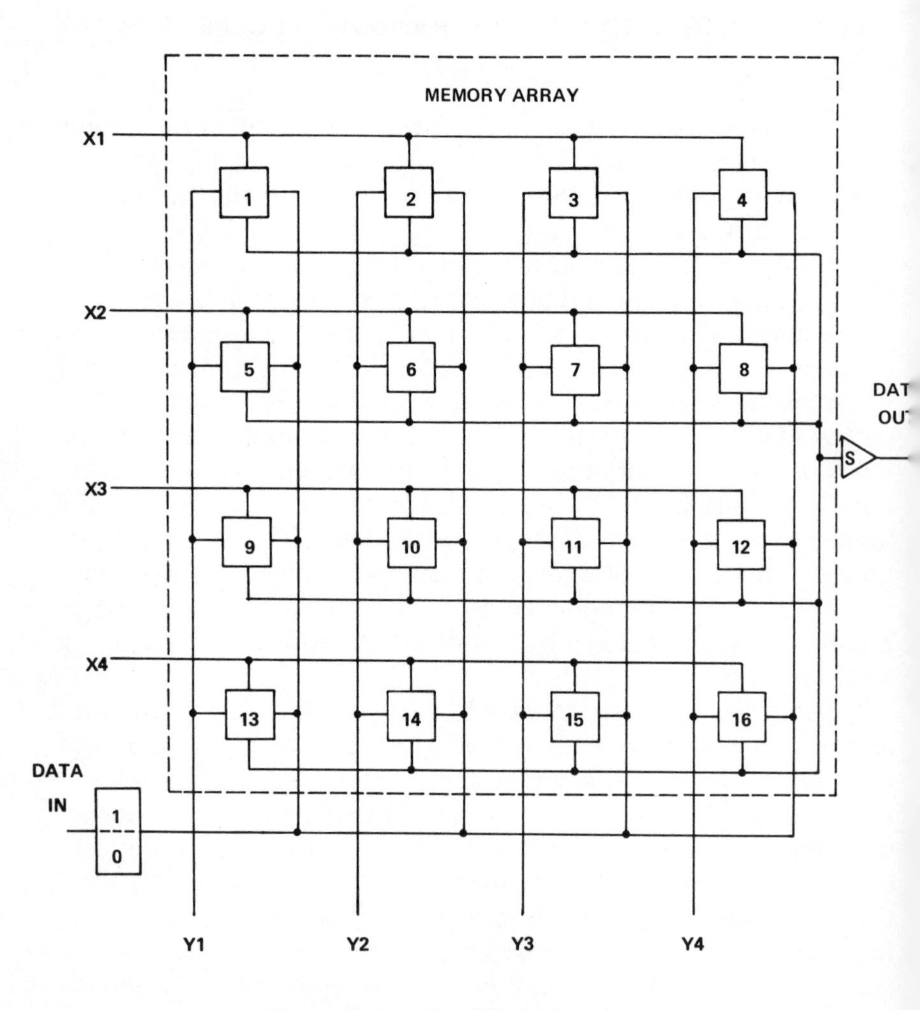

Figure 7-1. Simplified Drawing—16-bit Memory Array

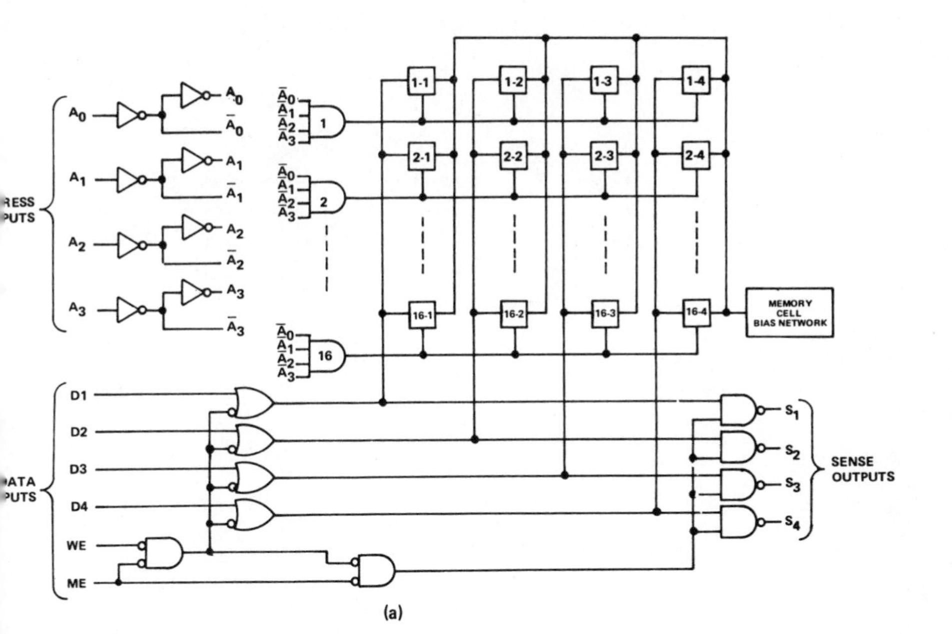

(a)

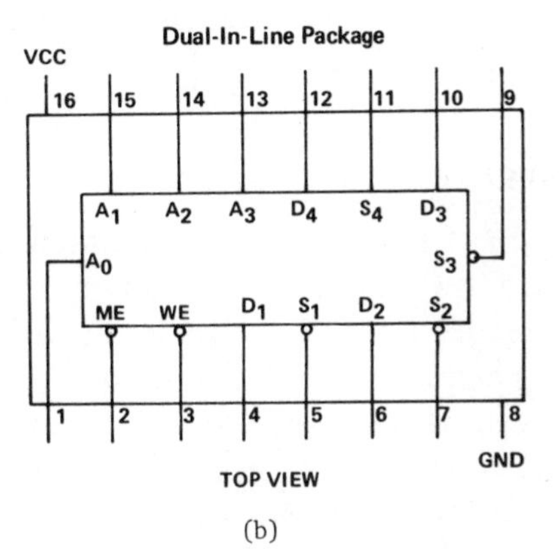

(b)

MEMORY ENABLE	WRITE ENABLE	OPERATION	OUTPUTS
0	0	Write	Logical "1" State
0	1	Read	Complement of Data Stored in Memory
1	X	Hold	Logical "1" State

(c)

Figure 7-2. 7489 64-bit Random Access Read/Write Memory (a) Logic Diagram (b) Pin Connections (c) Truth Table *(Courtesy National Semiconductor Corp.)*

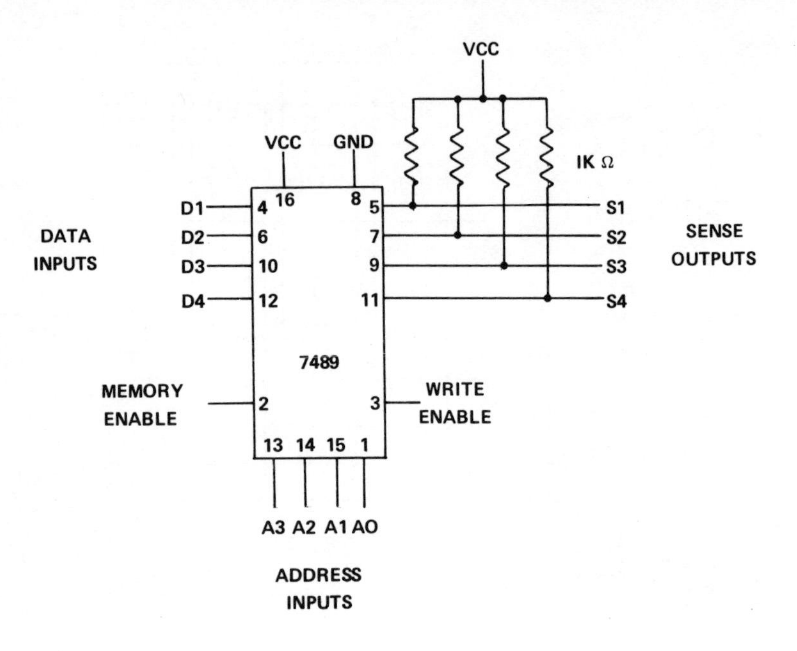

Figure 7-3. Pin Rearrangement and Pull-up Resistors

Increasing Word Length

By using two 7489 ICs, you can increase the word length from 4 bits to 8 bits as shown in Figure 7-4. The address lines of each IC are connected together respectively. The ME inputs and WE inputs are connected together. The pull-up resistors are used, but are not shown in the figure. The read/write operation is the same as for a single 7489 IC, except now you have 8 data inputs and 8 sense outputs, resulting in a 16-x 8-bit word RAM. Adding more 7489 ICs in this manner will increase the word length even more.

Increasing the Number of Words

You can increase the number of 4-bit words of the memory by connecting two 7489 ICs as shown in Figure 7-5. The data input lines and the sense output lines of each IC are connected together respectively. The WE inputs labeled WE-1 and WE-2 can be connected together or used separately. The addressing and read/write operations for this arrangement are slightly more complex than with the preceding arrangements. In order to read the first 16 locations, you have to address inputs A_0-A_3 and have input ME-2 high, while input

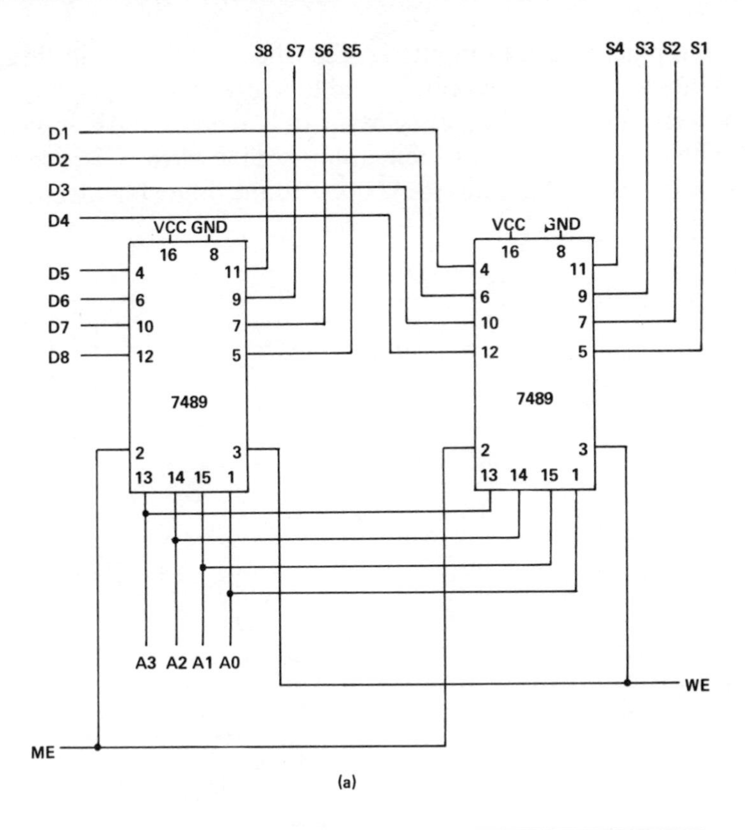

(a)

Operation	Address Locations	ME	WE	Ouputs S_1-S_8
Read	A_0-A_3	0	1	O's for stored data
Write	A_0-A_3	0	0	All 1's

(b)

Figure 7-4. Increasing Word Length (16x8-bit word RAM) (a) Wiring Diagram (b) Truth Table

ME-1 goes low to enable IC1. The stored data will appear at sense outputs S_1-S_4. To read the second 16 locations, you have to address inputs A_4-A_7 and have ME-1 high, while input ME-2 goes low to enable IC2. The stored data will also appear at sense outputs S_1-S_4. The write-in operation is similar, in that the desired IC must be selected by the proper ME input, addressed, and the desired data to be stored placed at data inputs D_1-D_4 when the WE input line goes low. Address inputs A_0-A_3 can be tied to address inputs A_4-A_7 respectively since the desired IC is selected only when its ME input goes low. The only precaution to remember is that when IC2 is selected, memory location 17 is addressed 0000 and the following locations ad-

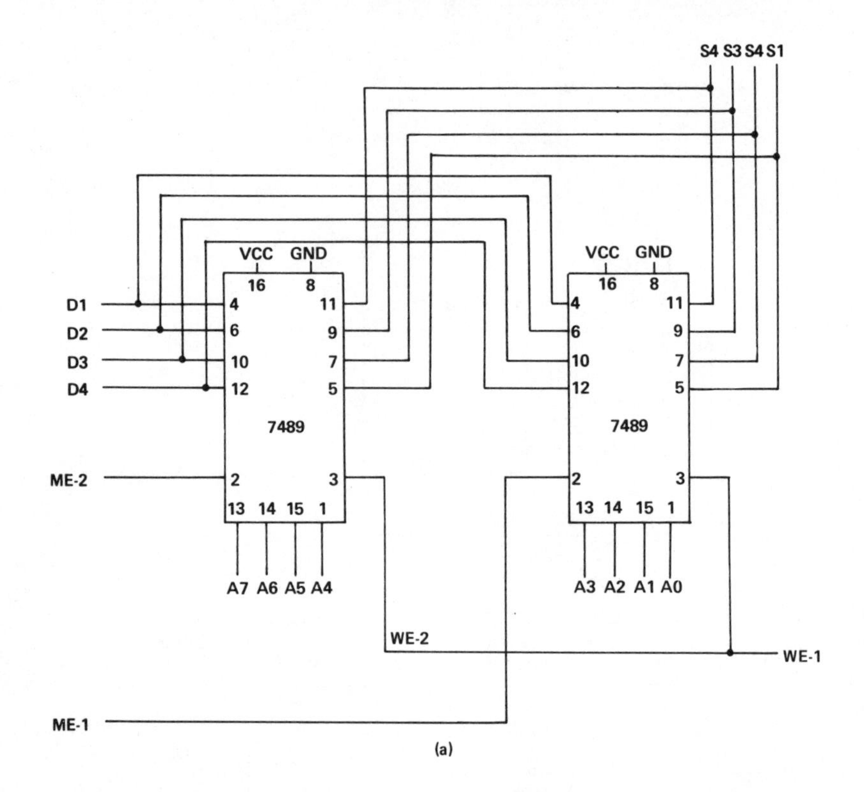

(a)

Operation	Address Locations	ME-1	ME-2	WE-1	WE-2	Outputs S_1-S_4
Read	A_0-A_3 A_4-A_7	0 1	1 0	1 1	1 1	O's for stored data
Write	A_0-A_3 A_4-A_7	0 1	1 0	0 0	0 0	All 1's

(b)

Figure 7-5. Increasing the Number of Words (32- x 4-bit word RAM) (a) Wiring Diagram (b) Truth Table

dressed respectively up to location 32 where the address inputs would be 1111.

Other 7489 ICs can be added to increase the number of words in the memory, whereas this arrangement will produce a 32- × 4-bit word RAM.

Other Types of RAMS

There are other types of RAMs which have larger storage capacity and/or special features that are produced in IC form. Some RAMS are organized as 4 × 4, 16 × 4, 256 × 1, 256 × 4, and 4096 × 1-bit word memories. Other memories are capable of simultaneous read/write operations, code converters and content addressable memory (CAM), where a simultaneous search of its entire contents is made to determine if it contains the data that is being presented. If the CAM already contains the desired data, the memory will not store it.

74170 4-by-4 Register File

The 74170 4-by-4 register file IC is actually a 4- × 4-bit word RAM. Although it has less bits than the 7489 RAM previously mentioned, it does have the unique feature of being able to write data into one location while reading data out of another location at the same time. The 74170 has two address select inputs as shown in Figure 7-6. The write select address (pins 13 and 14) are used in conjunction with the write enable input (pin 12) and the read select address (pins 4 and 5) are used with the read enable input (pin 11). The data inputs and outputs operate the same as the 7489, except the output appears in the true form rather than being complemented.

74200 256-bit RAM

The 74200 256-bit RAM is organized as a 256- × 1-bit word memory. This IC has the three-state (or Tri-State®) logic feature, which is discussed in Chapter 3. It has eight address inputs, one data input and output, three chip enable (memory enable) inputs, one write enable input and power connections as shown in Figure 7-7. The WE input determines the read mode or write more, while the three CE inputs determine whether the output is in the conventional true state, the complemented state or off condition.

The 7489 64-bit read/write RAM is rather limited. Instead of adding more 7489 ICs to increase the word length or number of words as we have previously done, it is more efficient from the standpoint of construction, operation and cost to use other RAMS with more capability. For instance, to construct a 256- × 8-bit word RAM, we would need thirty-two 7489 ICs designed with a complex wiring scheme that would require a difficult method of operation. Whereas, we could use eight 74200 ICs connected in parallel with a relatively simple means of operation. The cost of these memories at the time of the writing of this book would be $96.00 for the 7489 ICs as compared to $56.00 for the 74200 ICs.

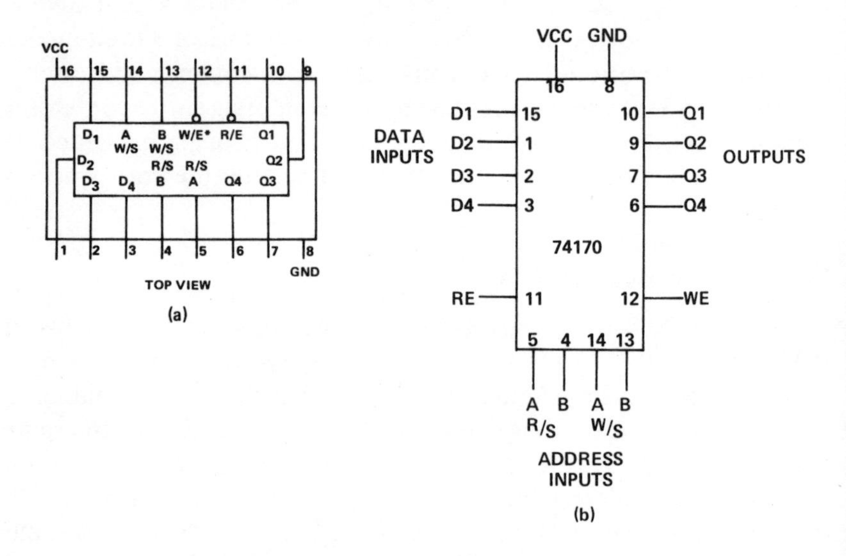

Figure 7-6. 74170 4-by-4 Register File IC (a) IC Pin Configuration *(Courtesy National Semiconductor Corp.)* (b) Pin Rearrangement

7-2 UNDERSTANDING THE READ-ONLY MEMORY (ROM)

The read-only memory or ROM is a memory from which data can be read out repeatedly, but cannot be written into. The basic ROM is programmed by the manufacturer with photo-masking techniques, while the IC chip is being fabricated. When the customer orders the ROM from the manufacturer, he also supplies the information for the type of data to be stored. An illustrative example of how the ROM is programmed is shown in Figure 7-8.

Only two 4-bit words of the memory array are shown in the figure. When word 1 is selected to be read, its base line goes high. The

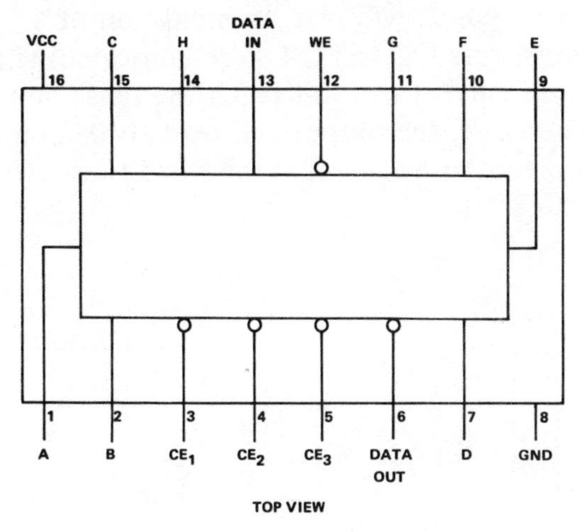

(a)

CE	WE	OPERATION	OUTPUT (DM8582)	OUTPUT (DM74200)
L	L	Write	Logical "1" (Open Collector)	High Z
L	H	Read	$\overline{D}$ (Complement of Stored Data)	$\overline{D}$ (Complement of Stored Data)
H	L	Do Nothing	Logical "1"	High Z
H	H	Do Nothing	Logical "1"	High Z

(b)

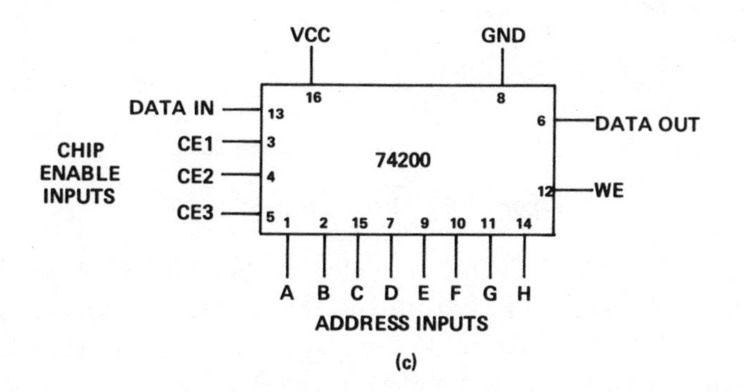

(c)

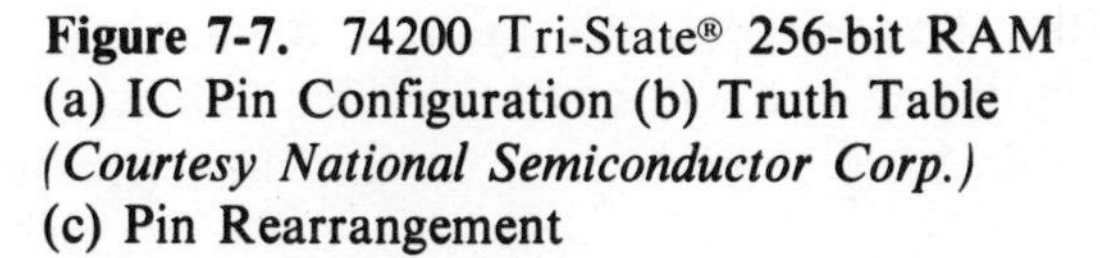

Figure 7-7. 74200 Tri-State® 256-bit RAM
(a) IC Pin Configuration (b) Truth Table
(Courtesy National Semiconductor Corp.)
(c) Pin Rearrangement

transistors will normally turn on and allow current to flow to the sense amplifiers, which will give an indication of a 1 at the output. However, transistors Q2 and Q4 were fabricated with their emitter leads open and current can not reach the sense amplifiers in these positions. Therefore, the output will read 1010. Transistors Q5 and Q7 have open emitter leads and when word 16 is selected the output will be 0101.

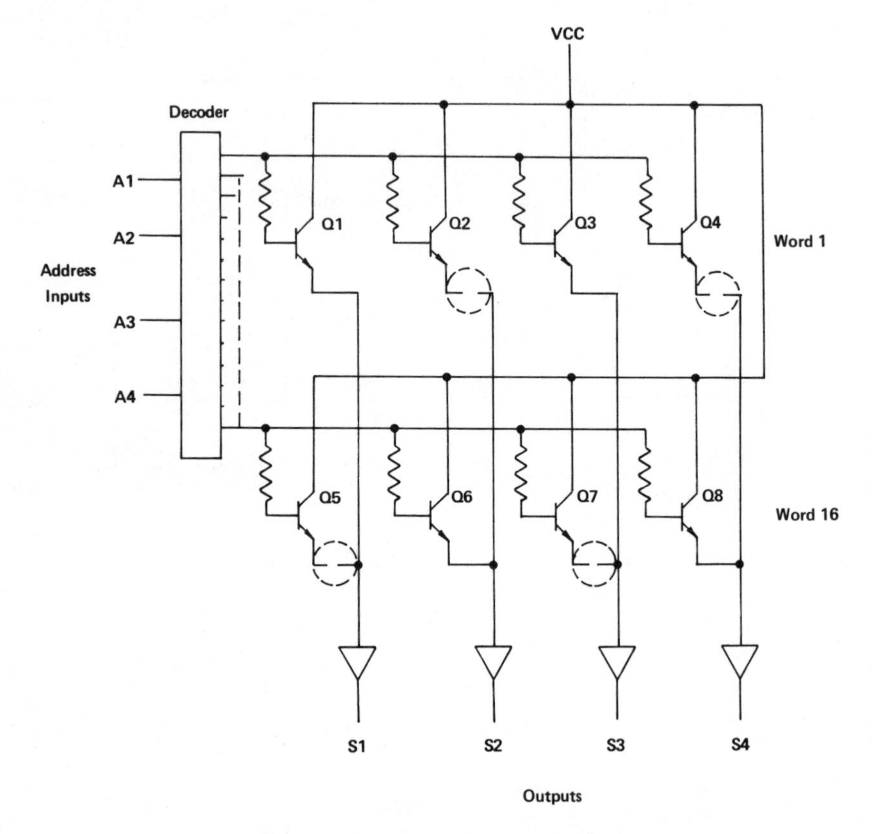

Figure 7-8. Programmed ROM

The 7488 256-bit ROM is one of the newest ROMs in the 7400-IC series and is shown in Figure 7-9. It is a custom-programmed 256-bit ROM organized as a 32- × 8-bit word memory. There are five address select inputs A-E and eight outputs Y1-Y8. The memory enable input overrides the address select inputs and when it is high all of the outputs are inhibited and will be at the logic 1 state. The outputs are of the open collector type permitting expansion to greater numbers of words.

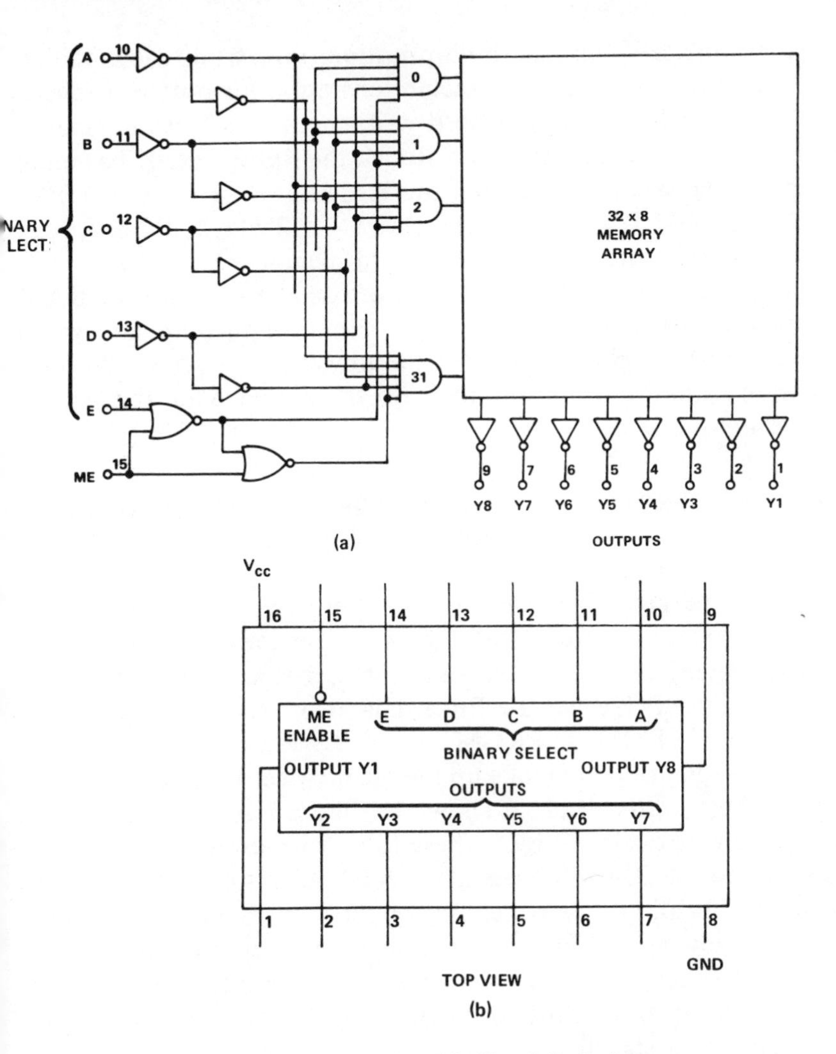

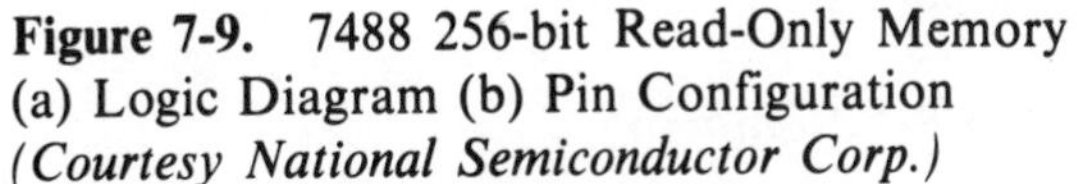

Figure 7-9. 7488 256-bit Read-Only Memory (a) Logic Diagram (b) Pin Configuration *(Courtesy National Semiconductor Corp.)*

Major Differences Between RAMs and ROMs

The RAM is desired when the data in the memory must be changed at various times. On the other hand, the ROM is usually a special circuit designed for a particular customer where only specific data needs to be read-out. Some information is useful in more than one application, hence a few standard ROMs have been developed. These standard ROMs may be programmed for code conversions, reference tables (such as multiplication tables, division tables, trigonometric tables), and character generation for output displays.

Since the ROM doesn't need write-in circuits and its memory cell is more simple than that of a RAM, the ROM usually has more bits per same size package than the RAM.

Some ROM ICs are organized as 32 × 8, 256 × 4, 512 × 4, 1024 × 4, 1024 × 8, and 2048 × 8 bit word memories.

The ROM is a non-volatile memory as compared to the RAM volatile memory. A volatile memory is a semiconductor memory that loses its stored digital data when the power is removed. The open emitter circuits in the ROM indicate the stored data and are, of course, non-volatile.

7-3 UNDERSTANDING THE PROGRAMMABLE READ-ONLY MEMORY (PROM)

The programmable read-only memory is one member of the standard semiconductor ROM family of memories. It differs only from the ROM in that the customer or user is able to program the memory himself. Hence, the PROM is also referred to as a field-programmable ROM or FROM.

The transistors that make up the memory cells are designed with fuses in their emitter leads as shown in Figure 7-10. The user is able to remove electrically or blow these fuses with external power supplies. A new PROM will have all of its ON-chip fuses intact and the output will be all 1s or 0s depending on the manufacturer. With some PROMs, blowing the fuse of a transistor will represent a stored 1, whereas in other PROMs it will represent a stored 0. Therefore, it is very important to have the specification sheet and programming instructions for a specific PROM IC.

How a PROM Is Programmed

In order to conform to the rules of logic, a truth table must first be constructed to correctly address and program the PROM. This same truth table is also used after the PROM is programmed so that the stored information can be correctly retrieved.

The address code selection depends on the digital system in which the PROM is to be used. The information to be retrieved from the PROM will depend on what it is used for, such as a micro-program for a computer or character generation for an output display. A programming set-up and partial truth table are illustrated in Figure 7-11. Again, this is only an example of how the programming of a PROM is accomplished and you should refer to the manufacturer's spec-sheet for a specific PROM IC.

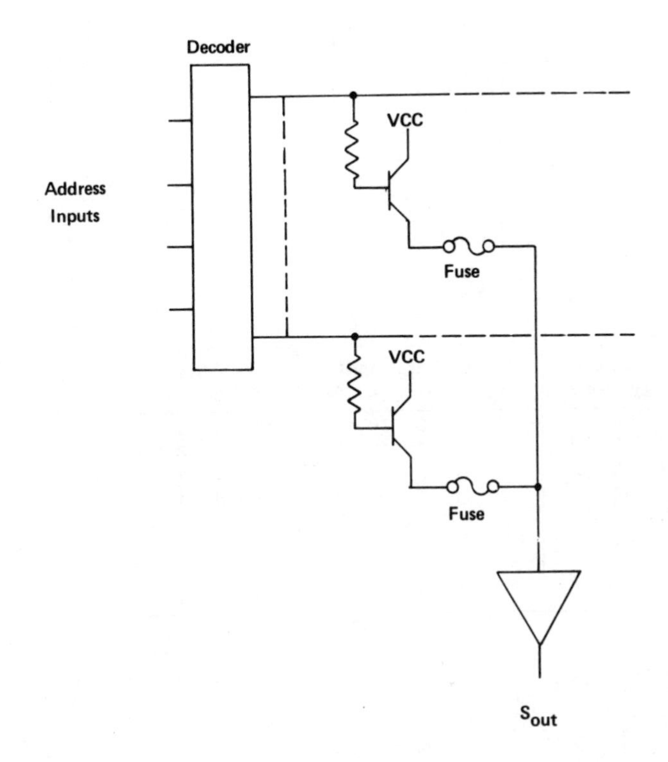

Figure 7-10. PROM on-chip Fuses

Logic switches are used to select the proper addresses. The memory enable input will usually be at the complementary logic level used to read out the stored data while the PROM is being programmed. A rotary switch SW1 is placed between the PROM outputs and a 12- to 25-volt d.c. regulated power supply. Switch SW2 is a normally open push-button switch in series with SW1 and the power supply. Some means of current limiting and indication should also be employed.

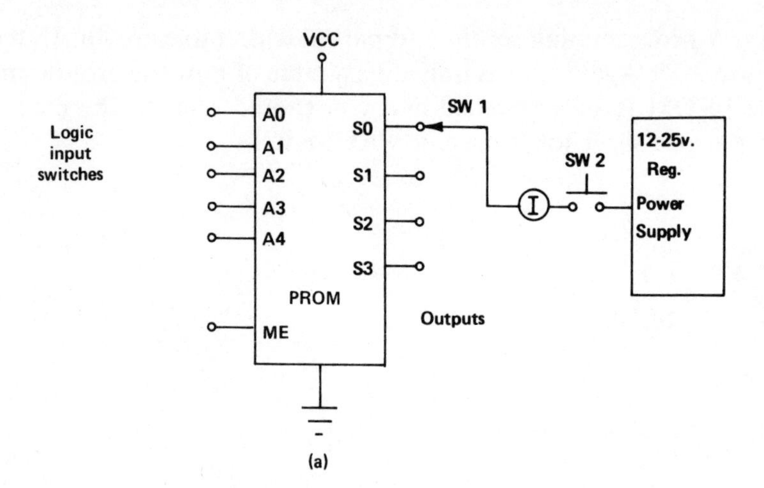

Program Sequence	Address Input					Outputs			
	A_4	A_3	A_2	A_1	A_0	S_3	S_2	S_1	S_0
1	0	0	0	0	0	0	0	1	1
2	0	0	0	0	1	0	1	0	1
3	0	0	0	1	0	0	1	1	1
4	0	0	0	1	1	1	0	1	0
•			•					•	
•			•					•	
•			•					•	
32	1	1	1	1	1	0	1	1	0

(b)

Figure 7-11. Programming a PROM
(a) Programming Connections (b) Truth Table

The actual programming procedure would go something like this:

1. Apply Vcc and ground to the PROM IC.
2. Select the desired address with the logic input switches. (Referring to the truth table, all of the switches would be low.)
3. Set SW1 to output S0.
4. Depress SW2 for about a half second and release. Observe the current meter and stay within the manufacturer's specifica-

tion. This action will blow the fuse of the selected transistor and in so doing a considerable amount of on-chip heat is developed.

5. Allow several seconds for the chip to cool down and then bring the ME input to the proper level for a read-out to check the fuse blowing operation. A logic probe would be fine for checking the output.
6. Set SW1 to output S1 and repeat steps 4 and 5.
7. The word 1100 is now programmed in address location 0000. Continue to select each address and program the bits of each word one at a time, according to the truth table, until the programming of the PROM is completed.

Some PROMs that are TTL compatible and are contained in a 16 pin dual-in-line package are:

Signetics—8223 256-Bit
Bipolar PROM (organized
32-× 8-bit word)

National Semiconductor
Corporation—8573 1024-Bit
PROM with open collector
(organized 256- × 4-bit word)
8574 TRI-State ® 1024-Bit
PROM (organized 256- × 4-bit word)

Erasable PROMs

The PROM is usually thought of as having permanent data stored in its memory. Newer PROMs now on the market are erasable and the data is considered to be semi-permanently stored.

One type of erasable PROM made by National Semiconductor Corp, the 2048-bit MM 5203, is encased in a 16 DIP with a quartz window on top that is transparent to short-wave ultraviolet light. The unwanted data is simply erased by exposing the window to ultraviolet light. The PROM can then be reprogrammed with the new data.

The memory element consists of a P-type MOSFET with floating silicon gate located between the source and drain. The gate has no external connection. During programming, a voltage in excess of —30 volts is applied across the drain and source, resulting in the injection of high-energy electrons into the floating silicon gate. This negative charge on the gate allows current to flow between source and drain during read-out.

To erase the stored data, the PROM is placed about an inch away from an ultraviolet light source for about 20 minutes. The erasing action results from an ionizing effect which causes the excess electrons on the floating gate to flow back into the substrate. The name given to this type of transistor is *floating gate avalanche-injection metal oxide semiconductor,* which is abbreviated, FAMOS.

Other types of erasable PROMs can be erased electrically and reprogrammed up to a million times.

7-4 MEMORY TERMINOLOGY

This section lists some of the more commonly used terms associated with memories and their uses. The terms are not in alphabetical order, but are grouped in relation to each other.

Memory: a device in electronics that can electrically, magnetically, or by other means store a 0 or 1 bit.

RAM (Random Access Memory): a memory where each location can be addressed directly and any particular data written into or read out without relation to any other locations.

Dynamic RAM: a random access memory that uses a charge-storage device for storing a bit.

Static RAM: a random access memory that uses flip-flops for storing bits.

ROM (Read Only Memory): a memory programmed by the manufacturer from which data can be read-out repeatedly, but cannot be written into.

PROM (Programmable Read Only Memory): a ROM that is programmed by the user rather than the manufacturer.

FROM (Field Programable Read Only Memory): same as *PROM*.

Memory Cell: the basic unit of storage within a memory such as a flip-flop or magnetic core.

Memory Array: The matrix or network of memory cells that consists of the actual storage of a memory.

Volatile Memory: a memory device in which the stored data is lost when the power is removed.

Non-Volatile Memory: A memory device in which the stored data is not lost when the power is removed.

Store: to place data in the memory.

Write-in: same as store.

Retrieve: to take data out of the memory and make it available to other circuits.

Read-out: same as retrieve.

Access Time: the time it takes for data to be written into or read out of a memory. More specifically, the read-time is access time required to retrieve data and the write-time is the access time needed to store data.

Address (noun): the specific location of stored data in a memory.

Address Register: the register which temporarily holds the information needed to select a specific address in the memory.

To Address (verb): the act of selecting a desired memory location for data to be written into or read out of the memory.

Sense Amplifier: an amplifier which senses the presence of a 1 or 0 bit stored in a memory. A four-bit word memory would require four sense amplifiers.

Memory Data Register: the register that temporarily holds the data that is to be written into or read out of the memory.

Chip Enable: an input to a memory IC that controls the read/write operation of the IC. The IC can be selected by this input.

Memory Enable: same as chip enable.

Program (noun): a set of stored instructions and data in digital form that controls the operation of a computer or digital device.

To Program (verb): the act of storing instructions and data into a memory device.

Byte: An 8-bit computer word which may contain bit subgrouping to facilitate information handling.

Microprogram: the most basic instructions that are "hardware" executed by a computer. The microprogram is frequently stored in a ROM whose outputs control various logic gates and circuits that perform the digital operations.

Inner Memory: the main memory in a computer where operating programs are temporarily stored.

Auxiliary Memory: other storage devices connected to the CPU which stores various programs and data.

Scratch Pad Memory: Small memories, usually ROMs, that perform sub-routines of a given program or store predefined mathematical tables and codes.

Semiconductor Memory: An electronic memory device using transistors, diodes, or flip-flops as memory cells.

Magnetic Core Storage: a memory which uses a small round magnetic "doughnut" shaped core material with wires around it or going through it for magnetically storing data.

Magnetic Drum: A rapidly rotating cylinder whose surface can be easily magnetized for storing data.

Magnetic Disc: a constantly rotating flat circular plate (similar to an LP phonograph record) with magnetic surfaces which can store data. Read/write sense recording heads are positioned over the surface of the disc.

Magnetic Tape: a magnetically-coated plastic tape used to store data, similar to standard reel-to-reel tape recorders.

Thin-Film Memory: a magnetic memory that is fabricated similar to semiconductor devices.

Punched Cards: cardboard cards with holes punched in various columns representing stored 1 bits.

Punched Paper Tape: a strip of paper tape about one-inch wide with punched holes representing the stored 1 bits.

SUMMARY

Semiconductor IC memories basically consist of RAMs, ROMs, PROMs and erasable PROMs. The user can write in or read out data rather easily with a RAM. A ROM is usually programmed by the manufacturer according to the user's needs and can only be read out. The PROM is a ROM which is programmed in the field by the user. Erasable PROMs are useful if a mistake is made during the programming operation or if it is desired to change the stored data.

Single IC memory packages can be connected together to increase the length of a word and/or increase the number of words in a memory system.

The RAM is a volatile memory, while the ROM is a non-volatile memory which is very useful for the microprogramming of basic instructions, code conversions, fixed tables and character generation.

7-5 PROJECT # 7: WIRING AND TESTING A 64-BIT READ/WRITE MEMORY

In Project # 7, you will be able to store any sixteen 4-bit digital words of your choice and retrieve them at will. At the same time you will be able to test all of the memory cells in the array. The 7489 RAM IC will be used in a later project to show how character generation is accomplished with an output display.

In addition to the breadboard/tester and connecting wires used in the previous projects, you will need four 1K-ohm pull-up resistors for the outputs of the 7489 IC.

It may be advantageous to build another smaller auxiliary breadboard with IC sockets and pull-up resistors as shown in Figure 7-12. This breadboard can also be used with the output displays described in Chapter 9.

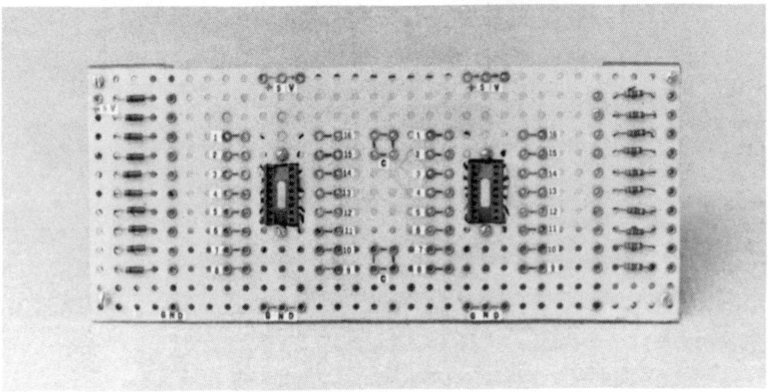

Figure 7-12. Auxiliary Breadboard

The row of ten 1K-ohm resistors on the left side of the breadboard is used for IC pull-up resistors. The row of ten 330-ohm resistors on the right side of the breadboard is used for the output display devices. The left side of the 1K-ohm resistors is permanently connected together and is tied to the +5V input terminal. The other

+5V terminals are also common to this +5V input terminal. All of the ground (GND) terminals are permanently connected together. Each of the two pairs of four terminals between the IC sockets marked C is common and is used for connecting common wiring points when needed. The 330-ohm resistors have each end connected to a separate terminal to be used in later projects.

In order to test the 7489 RAM IC the wiring can be done as shown in Figure 7-13. Remember, that since the output of the 7489 RAM IC is the complement of what is stored at the input, a 7404 hex inverter is needed to drive the LEDs.

Input switches A-D are used for the data inputs while input switches E-H are used for the address select inputs. Input switch I is usually in the low position and serves as the memory enable. The "0" trigger pulse is used to write data into the memory.

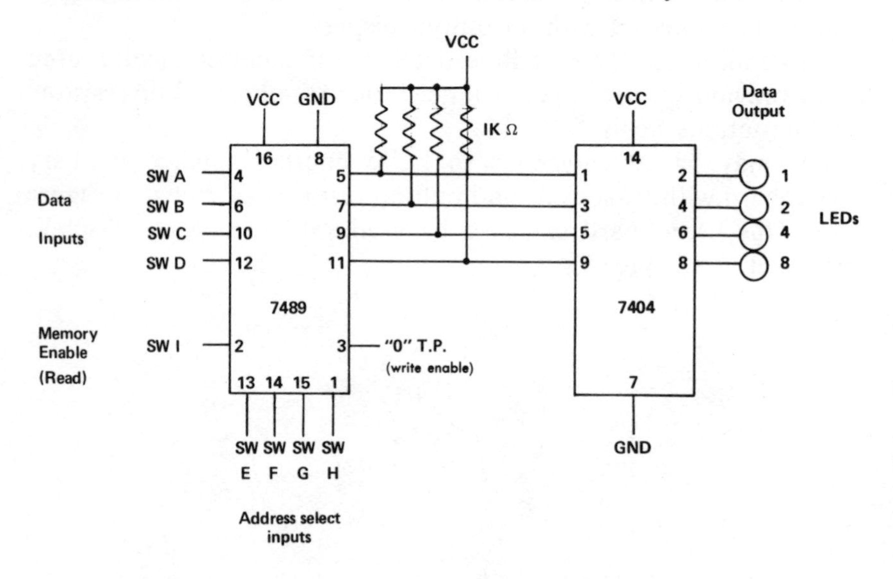

Figure 7-13. Wiring Diagram for testing the 7489 64-Bit Read/Write Memory IC.

Testing Procedure for the 7489 RAM IC

You should first of all make up a truth table and decide what binary data you want to store at each of the 16 locations of the memory.

To write-in and read-out binary data:

1. Select the desired address by setting switches E-H to proper position.
2. Place desired binary data at the data inputs by setting switches A-D to the proper position.
3. Make sure switch I is low.
4. Momentarily press the "0" T. P. switch. The stored data will immediately appear at the LED outputs.
5. Place all data input switches A-D to a low and again check the LED outputs.
6. Momentarily place switch I to a 1 condition and then back to 0 condition. The LEDs will extinguish and then turn on again, indicating that the binary data is still stored at that address.

Follow this same procedure for the other 15 memory locations. *Caution!* Remember not to lose power to the 7489 RAM IC since it is a volatile memory and the stored data will be destroyed or altered.

Using two 7489 RAM ICs, you may want to increase the word length of the memory as shown in Figure 7-4 or increase the number of words in the memory as shown in Figure 7-5.

CHAPTER 8

Using Conversion and Switching Devices in Digital Systems

Generally, conversion devices are used to match two or more different systems. Since we deal with the base ten number system and digital circuits operate with the base two number system, some conversion devices are needed for us to communicate efficiently with digital circuits.

Encoders usually are used at the input terminals of digital systems for entering data, while decoders usually are used at the output terminals with readout devices to obtain the desired information. Decoders are also used within the digital system to execute commands and set up specific operations.

The IC memories described in the last chapter are used in the conversion process where data is actually stored in the device. Basic encoders and decoders are active logic networks that do not store data, but respond to the input data directly.

Multiplexers and demultiplexers are used to combine data lines from one circuit or digital system to another and considerably reduce the wiring needed between the various units in a digital system.

In this chapter you will learn how to construct and use encoders, decoders, multiplexers and demultiplexers, while **Project # 8** will give you experience in wiring and testing these devices.

8-1 ENCODERS

Encoders are usually considered as devices that convert other number systems into binary number systems. A simple decimal-to-binary encoder using OR gates is shown in Figure 8-1.

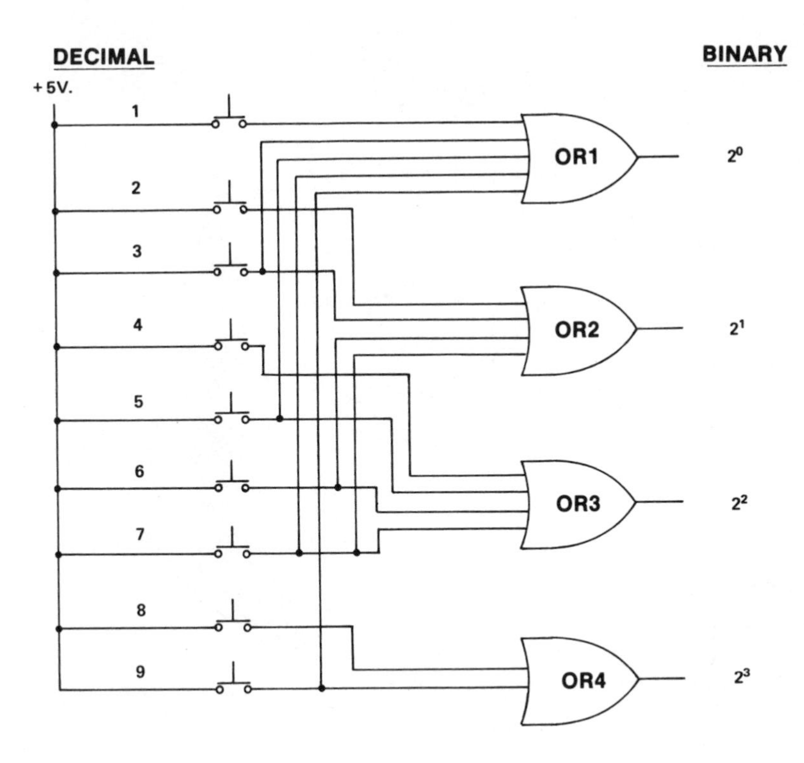

Figure 8-1. Decimal to Binary OR Gate Encoder

The left side of the switches is connected to a 1 or positive voltage source. When any decimal switch is pressed, one or more OR gates turn on, giving the binary equivalent output. For example, when switch 5 is pressed, gates OR1 and OR3 turn on and the binary output will read 0101. The other decimal switches can be traced to determine which OR gates turn on to represent the binary equivalent.

NAND gates can be used to construct an encoder as shown in Figure 8-2. The decimal switches should normally be high. When a switch is pressed, a low is applied to the corresponding gates which turn on and give the binary equivalent. If the switches give a high when pressed, inverters could be used as inputs to the NAND gates.

Another means of conversion in digital circuits, that may be less costly, is with a diode matrix. One type of diode matrix encoder is

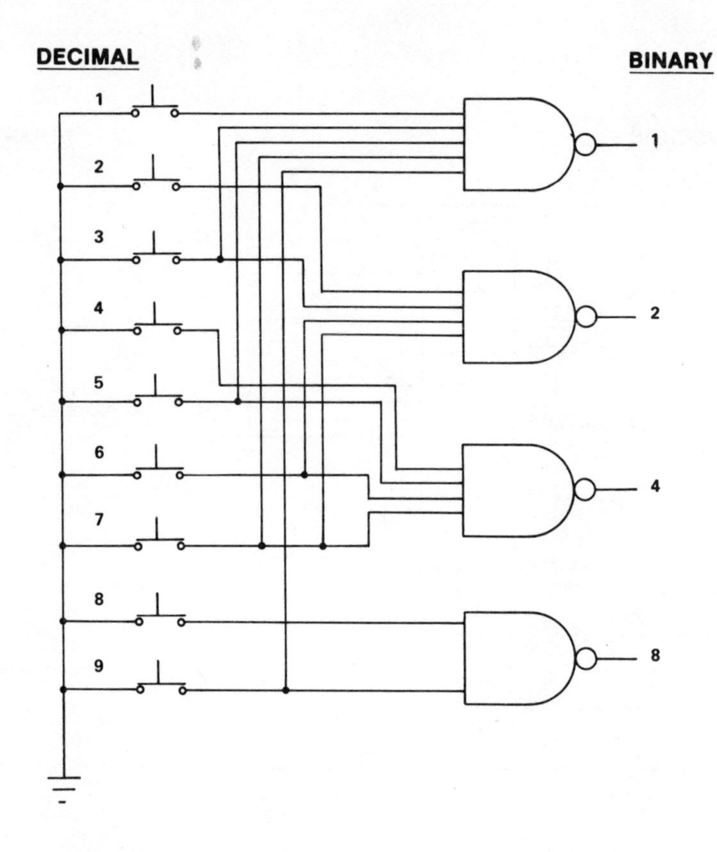

Figure 8-2. Decimal to Binary NAND Gate encoder

shown in Figure 8-3. The diodes and resistors are so arranged that when a decimal switch is pressed the diodes conduct, causing voltage drops to appear across the respective resistors.

As an example, assume decimal 5 switch is pushed and line 5 has a positive voltage applied to it. The four diodes with their anodes connected to line 5 will conduct:

Current flows from ground through R1, through D1 to line 5
Current flows from ground through R4, through D2 to line 5
Current flows from ground through R5, through D3 to line 5
Current flows from ground through R8, through D4 to line 5

The voltage drops of R1, R4, R5 and R8 will also appear at the respective binary outputs as indicated by the positive signs.

The binary output may be connected to a register, a counter, or other logic networks; therefore, the 0 is also represented as a pulse. There are eight binary outputs for all possible conditions.

Reading from left to right in the binary output, there is a 0 pulse, a 1 pulse, a 0 pulse and a 1 pulse indicating the binary number 0101.

This is a basic matrix and many variations are used to meet specific complex encoding and decoding requirements within digital circuits.

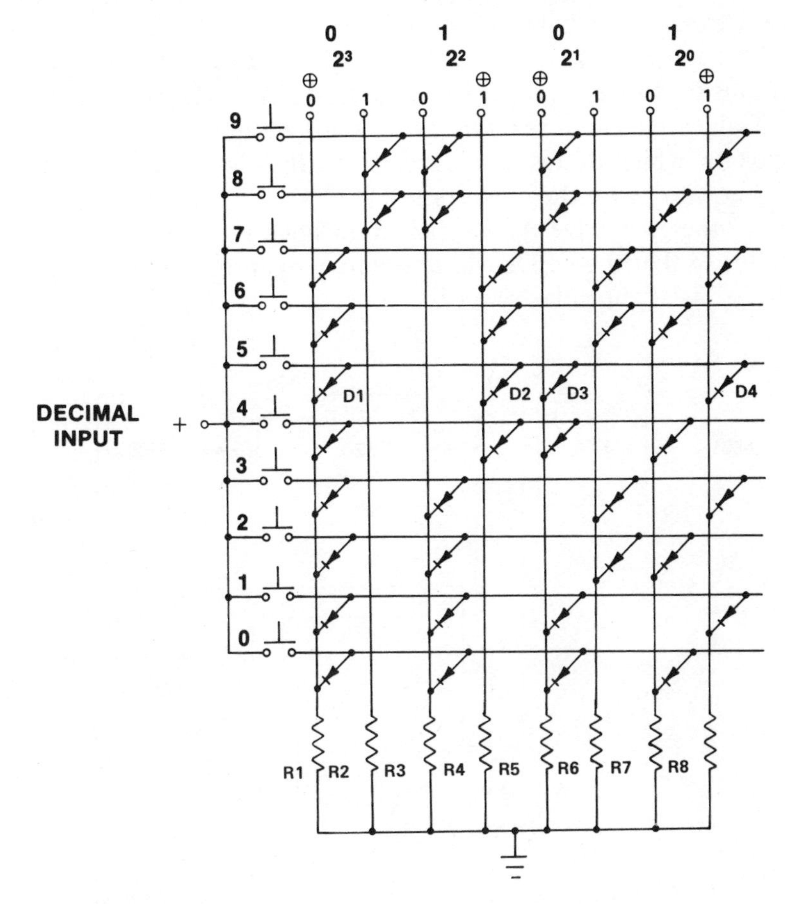

Figure 8-3. Decimal to Binary Diode Matrix Encoder

The encoders presented thus far all have had 4-bit outputs. When more bits per output are desired, the encoder may become very complex and require more physical area. Therefore, very often, ROMs are used as encoding devices.

8-2 DECODERS

Decoders are usually thought of as devices that convert binary number systems into other number systems and/or execute operations. The 4-bit switch-tail ring counter shown in Figure 5-6 used AND gates to decode the Q and $\overline{Q}$ outputs of the flip-flops and provide eight separate output pulses.

You may want specific action to occur in a given chain of events. As an example, Figure 8-4 shows two AND gates connected to a mod-8 counter. One AND gate will light its LED when the counter contains the binary number 011 and the other AND gate will light its LED when the counter contains the binary number 110. You can first decide in which states (on or off) the flip-flops will be for each number, then wire the corresponding Q and $\overline{Q}$ outputs of each flip-flop to the respective AND gate. You have now decoded a binary counter so that the LEDs will alternately light on the third and sixth pulse of an eight-pulse sequence.

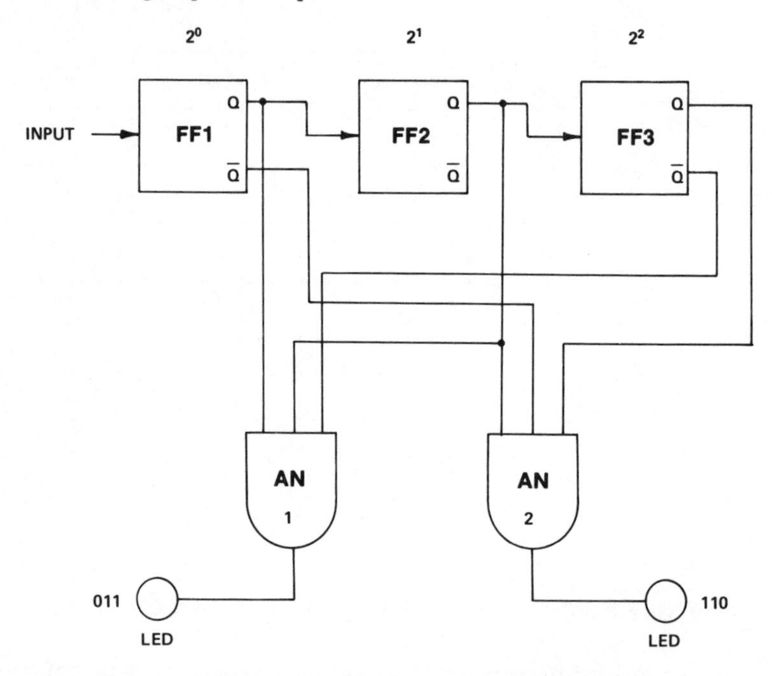

Figure 8-4. AND Gate Decoder

Decoders can be used to select specific input conditions as shown in Figure 8-5. The decoder network at the top will give a 1 output if two or more inputs are high. The various combinations are shown in the truth table. The decoder network at the bottom will give a 1 output if any input does not coincide with the other two inputs as shown in its truth table.

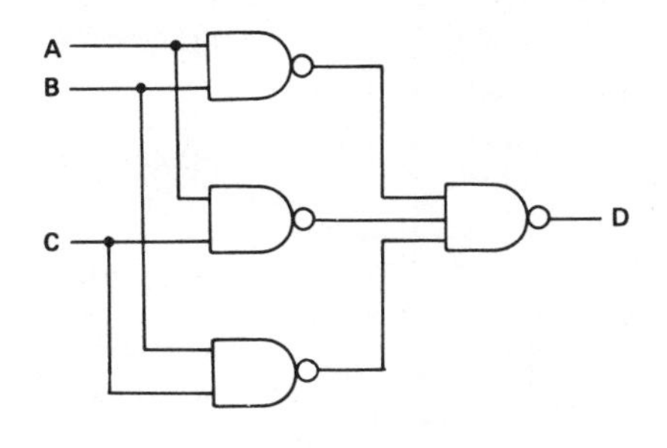

A	B	C	D
0	0	0	0
0	0	1	0
0	1	0	0
0	1	1	1
1	0	0	0
1	0	1	1
1	1	0	1
1	1	1	1

(a)

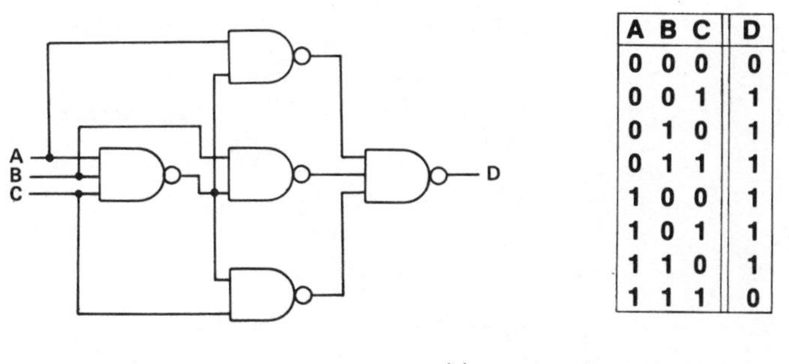

A	B	C	D
0	0	0	0
0	0	1	1
0	1	0	1
0	1	1	1
1	0	0	1
1	0	1	1
1	1	0	1
1	1	1	0

(b)

Figure 8-5. NAND Gate Decoder Networks
(a) Two or more inputs high
(b) NON-coincidence of one input

8-3 BCD-TO-DECIMAL DECODER

Multiple output decoders are needed for converting number systems, and some of these decoders have been standardized in IC form.

The 7442 BCD-to-Decimal Decoder IC is shown in Figure 8-6 and is often referred to as a one-out-of-ten decoder.

This IC accepts a 1-2-4-8 BCD number at inputs A-B-C-D

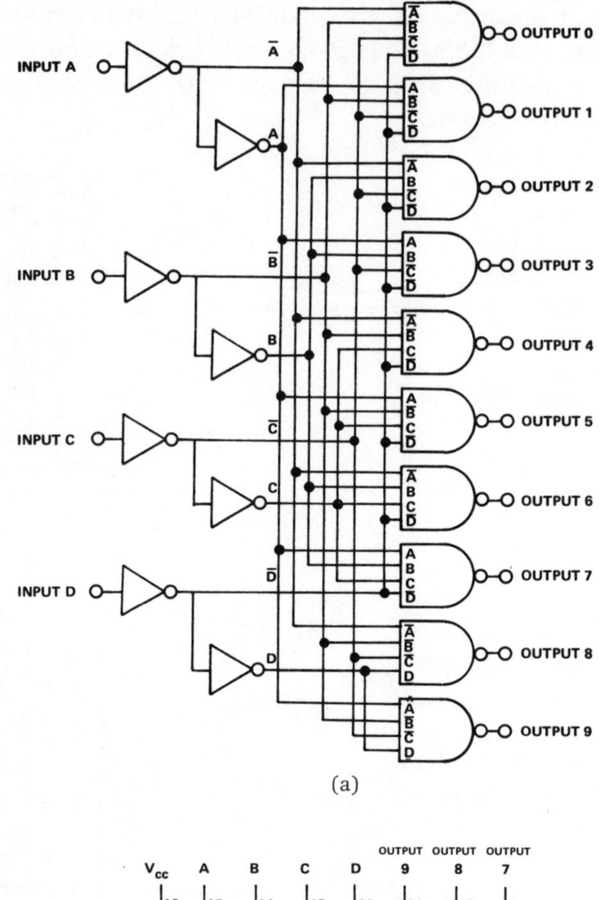

(a)

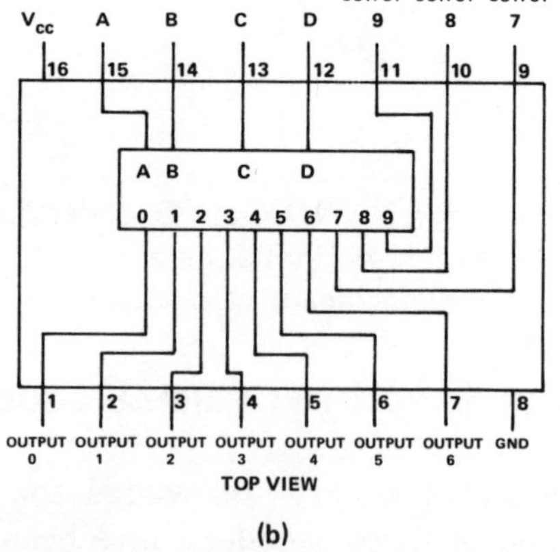

(b)

Figure 8-6. 7442 BCD-to-Decimal Decoder IC
(a) Logic Diagram (b) Pin Configuration (c) Truth Table
(Courtesy National Semiconductor Corporation)

INPUTS				OUTPUTS									
D	C	B	A	0	1	2	3	4	5	6	7	8	9
0	0	0	0	0	1	1	1	1	1	1	1	1	1
0	0	0	1	1	0	1	1	1	1	1	1	1	1
0	0	1	0	1	1	0	1	1	1	1	1	1	1
0	0	1	1	1	1	1	0	1	1	1	1	1	1
0	1	0	0	1	1	1	1	0	1	1	1	1	1
0	1	0	1	1	1	1	1	1	0	1	1	1	1
0	1	1	0	1	1	1	1	1	1	0	1	1	1
0	1	1	1	1	1	1	1	1	1	1	0	1	1
1	0	0	0	1	1	1	1	1	1	1	1	0	1
1	0	0	1	1	1	1	1	1	1	1	1	1	0
1	0	1	0	1	1	1	1	1	1	1	1	1	1
1	0	1	1	1	1	1	1	1	1	1	1	1	1
1	1	0	0	1	1	1	1	1	1	1	1	1	1
1	1	0	1	1	1	1	1	1	1	1	1	1	1
1	1	1	0	1	1	1	1	1	1	1	1	1	1
1	1	1	1	1	1	1	1	1	1	1	1	1	1

(c)

Figure 8-6. *(Continued)*

respectively, and provides a low at one of the ten outputs, while the other nine outputs remain high. For example, a 0101 input results in output 5 (pin 6) going low. Binary numbers from 10-15 at the input send all outputs high. Each output is capable of driving 10 standard TTL loads and can sink about 16 ma.

The 7445 BCD-to-decimal decoder/driver IC is identical in pin connections and operation to the 7442 IC, except that each output transistor is capable of sinking 80 ma and in the off condition, each transistor can withstand a breakdown voltage up to 30 volts.

The 7441 BCD-to-decimal decoder/nixie* driver is similar in operation to the 7442 and 7445 ICs, but its pin connections are different and it is designed to drive gas-filled-readout (NIXIE*) tubes or other low current lamps and relays. It also has an over-range feature that provides a decoded output if binary numbers between 10 and 15 are applied to the input as shown in Figure 8-7.

8-4 BCD-TO-SEVEN-SEGMENT DECODER/DRIVER

The seven-segment displays that are covered in the next chapter need a decoder in order to operate correctly. There are three standard BCD-to-seven-segment decoder/driver ICs. All of them have the same pin configurations as shown in Figure 8-8.

The 1-2-4-8 BCD input is applied to I_A (pin 7), I_B (pin 1), I_C (pin 2) and I_D (pin 6) respectively. The seven outputs appear on pins 9

*Trademark of Burroughs Corporation

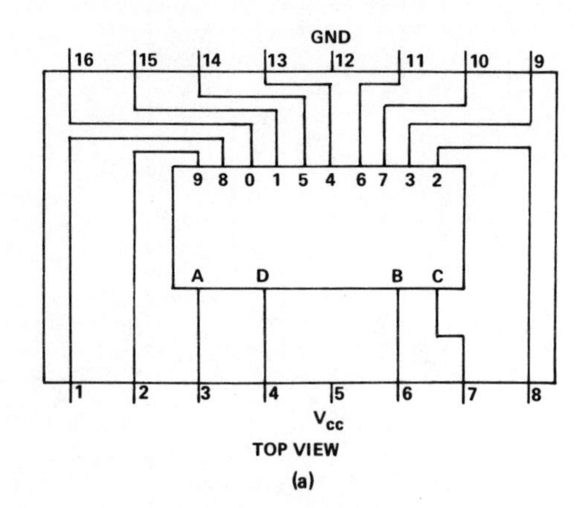

(a)

INPUT D	C	B	A	LOW OUTPUT
0	0	0	0	0
0	0	0	1	1
0	0	1	0	2
0	0	1	1	3
0	1	0	0	4
0	1	0	1	5
0	1	1	0	6
0	1	1	1	7
1	0	0	0	8
1	0	0	1	9
(OVER RANGE)				
1	0	1	0	0
1	0	1	1	1
1	1	0	0	2
1	1	0	1	3
1	1	1	0	4
1	1	1	1	5

(b)

Figure 8-7. 7441 BCD-to-Decimal Decoder/NIXIE* Driver (a) Pin Configuration (b) Truth Table *(Courtesy National Semiconductor Corporation)*

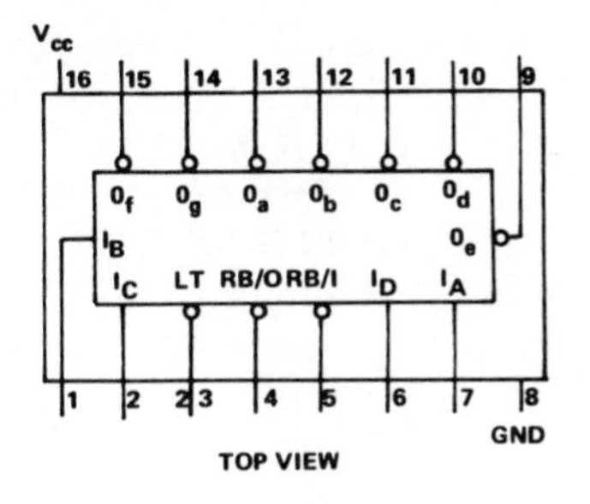

Figure 8-8. Pin Configuration for the 7446, 7447 and 7448 BCD-to-7-segment Decoder/Driver ICs *(Courtesy National Semiconductor Corporation)*

through 15 and are labeled $O_A - O_G$ as shown. Pin 3, labeled LT, is a lamp test input which will enable all seven outputs simultaneously, when brought low. The blanking input pin 5, labeled RB/I, will extinguish the character "0" when brought low and is used for turning off insignificant zeros in a multiple digit display. The blanking output, pin 4 labeled RB/0, will be low when RB/I goes low and this pin can also extinguish the entire output when brought to ground. The 7446 and 7447 have active-low open-collector outputs and will drive display segments requiring up to 40 ma of current. These ICs require a current limiting resistor of about 330 ohms when used with an LED display. They are capable of driving incandescent or fluorescent displays directly. The difference between these ICs is that the 7446 can withstand 30v at the output when high, whereas the 7447 can only handle up to 15 v. A truth table for these ICs is shown in Figure 8-9.

Decimal Numeral or Symbol	BCD Inputs D C B A	7-Segment Outputs a b c d e f g
0	0 0 0 0	0 0 0 0 0 0 1
1	0 0 0 1	1 0 0 1 1 1 1
2	0 0 1 0	0 0 1 0 0 1 0
3	0 0 1 1	0 0 0 0 1 1 0
4	0 1 0 0	1 0 0 1 1 0 0
5	0 1 0 1	0 1 0 0 1 0 0
6	0 1 1 0	1 1 0 0 0 0 0
7	0 1 1 1	0 0 0 1 1 1 1
8	1 0 0 0	0 0 0 0 0 0 0
9	1 0 0 1	0 0 0 1 1 0 0

Decimal Numeral or Symbol	BCD Inputs D C B A	7-Segment Outputs a b c d e f g
⊏	1 0 1 0	1 1 1 0 0 1 0
⊐	1 0 1 1	1 1 0 0 1 1 0
⊔	1 1 0 0	1 0 1 1 1 0 0
C	1 1 0 1	0 1 1 0 1 0 0
E	1 1 1 0	1 1 1 0 0 0 0
Blank	1 1 1 1	1 1 1 1 1 1 1

Figure 8-9. Truth Table for 7446 and 7447 BCD to 7-segment Decoder/Driver IC's. (The output is complemented with the 7448 IC.) *Note: See Figure 9-4e for LED seven-segment display.

The 7448 IC has active-high outputs and needs pull-up resistors of about 1000 ohms for proper operation. The source current is about 2 ma and the sink capability is 6.4 ma. It is normally used to drive logic circuits which operate high-voltage loads through buffer circuits. This IC requires the same BCD input as the 7446 and 7447 ICs, except that its output is complemented.

8-5 MULTIPLEXERS

A basic multiplexer is a switching device that can select one of a number of inputs and connect it to a single common output as shown in Figure 8-10.

The mechanical switch is, of course, selected by hand to connect any one of the input lines A-D to output line X. The digital mul-

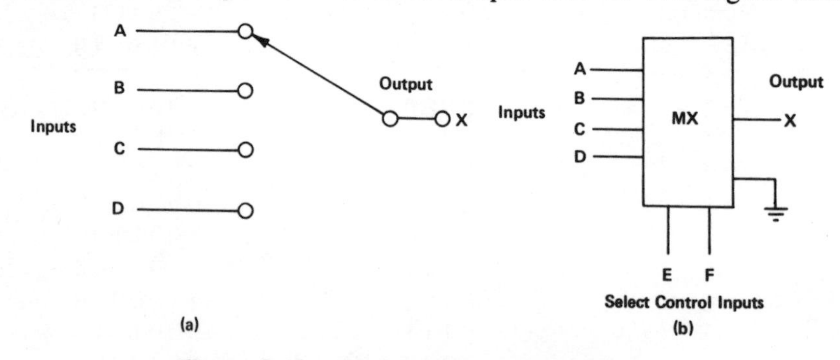

Figure 8-10. Multiplexers (a) Single-pole 4-position Switch (b) Digital Multiplexer

tiplexer will perform the same operation with a binary code applied to the select control inputs. The select control inputs work as a decoder which allows only one input line to be selected and pass through to the output as shown with the basic digital multiplexer in Figure 8-11.

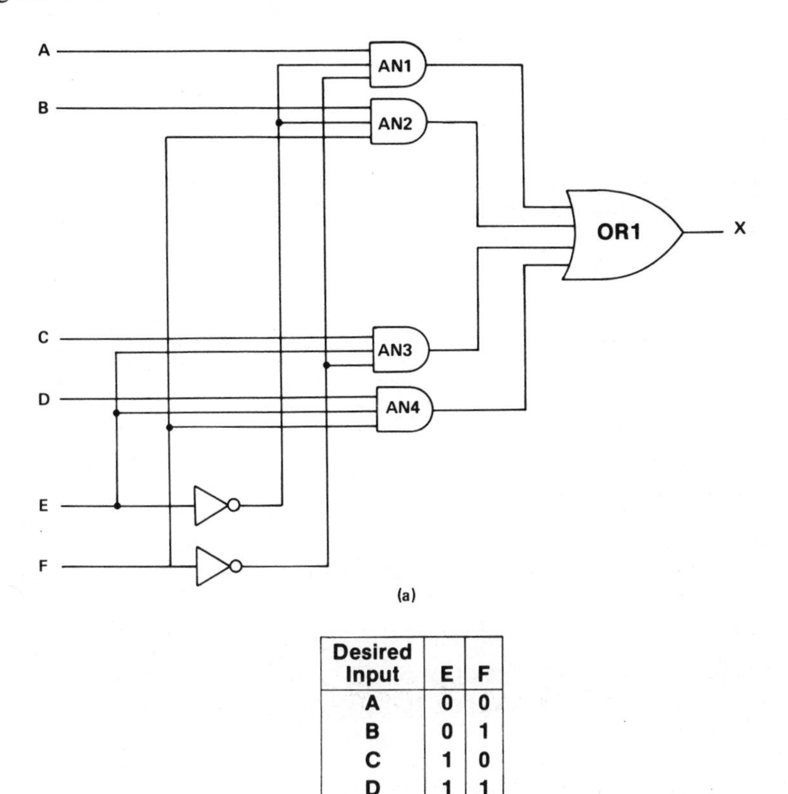

Desired Input	E	F
A	0	0
B	0	1
C	1	0
D	1	1

(b)

Figure 8-11. Basic Digital Multiplexer
(a) Logic Diagram (b) Input Select Truth Table

As an example, when lines E and F are low, input A is allowed to pass through gates AN1 and OR1 to the output X. The output will respond directly to the data at input A. The appropriate binary code will select any one of the other three inputs as shown with the input select truth table.

To construct this basic digital multiplexer, you could use four basic logic gate ICs, one 7404 hex inverter IC, two 7411 triple 3-input AND gate ICs and one 7432 Quad 2-input OR gate IC connected as a

4-input OR gate. This circuit could be implemented with NAND gates and only use three basic logic gate ICs, two 7410 triple 3-input NAND gate ICs and one 7420 dual 4-input NAND gate IC as shown in Figure 8-12.

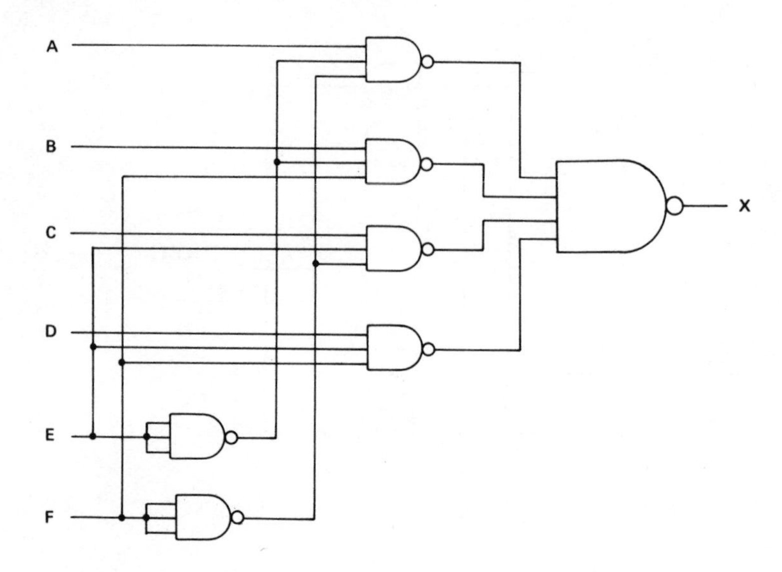

Figure 8-12. Implemented NAND Gate Multiplexer

Multiplexers are fabricated into a single IC as shown in Figure 8-13. The 74153 dual 4:1 multiplexer is similar to the basic multiplexer shown in Figure 8-11, but can also serve as a double-pole 4-throw

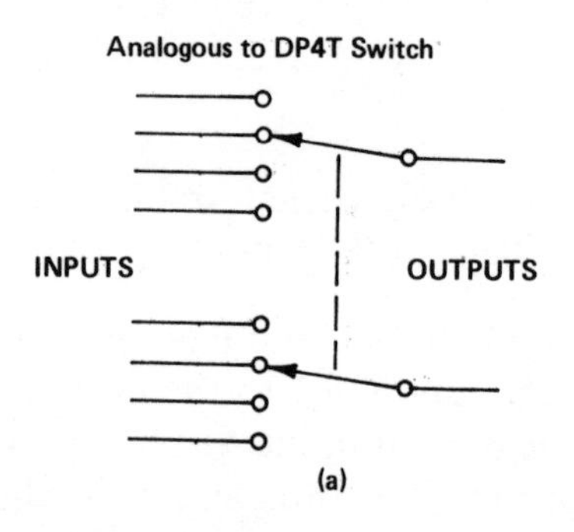

Figure 8-13. 74153 Dual 4:1 Multiplexer IC
(a) Double-pole 4-toggle Switch (b) Logic Diagram
(c) Pin Configurations (d) Truth Table
(Courtesy National Semiconductor Corporation)

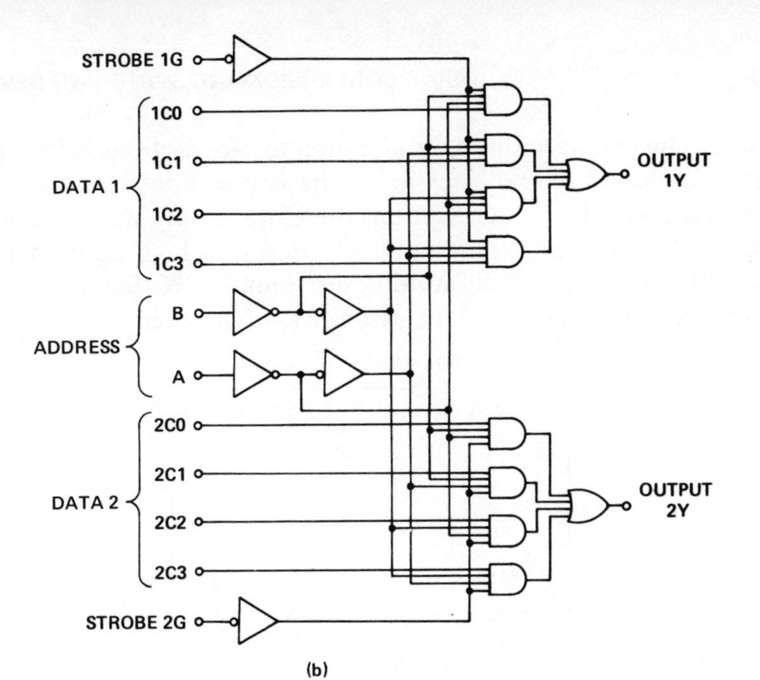

(b)

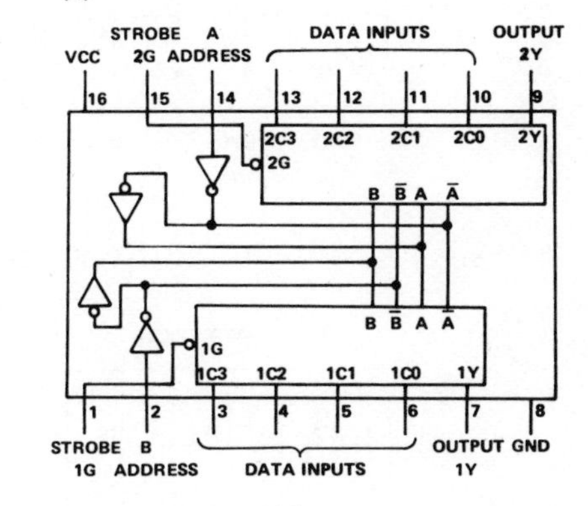

(c)

ADDRESS INPUTS		DATA INPUTS				STROBE	OUTPUT
B	A	C0	C1	C2	C3	G	Y
X	X	X	X	X	X	1	0
0	0	0	X	X	X	0	0
0	0	1	X	X	X	0	1
0	1	X	0	X	X	0	0
0	1	X	1	X	X	0	1
1	0	X	X	0	X	0	0
1	0	X	X	1	X	0	1
1	1	X	X	X	0	0	0
1	1	X	X	X	1	0	1

X = DON'T CARE

(d)

Figure 8-13. *(Continued)*

switch. The two select lines are common to each section and the same input is selected for both sections. The strobe input to each section provides control for each section, therefore if only one section was needed, one of these lines could be wired to a high permanently.

Other fairly common multiplexers found in IC form are shown in Figure 8-14. The 74151 IC is an 8:1 multiplexer with output Y

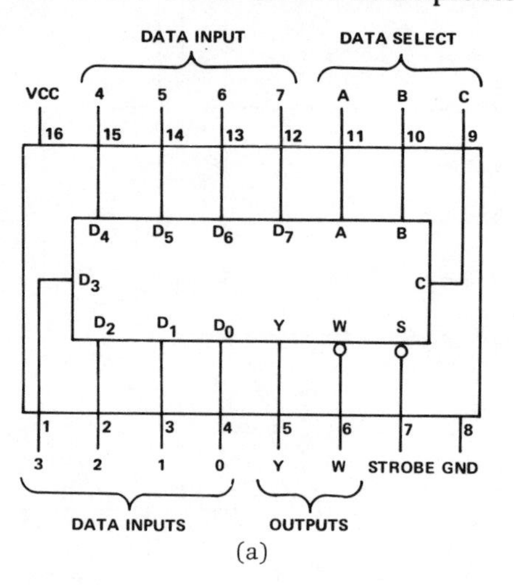

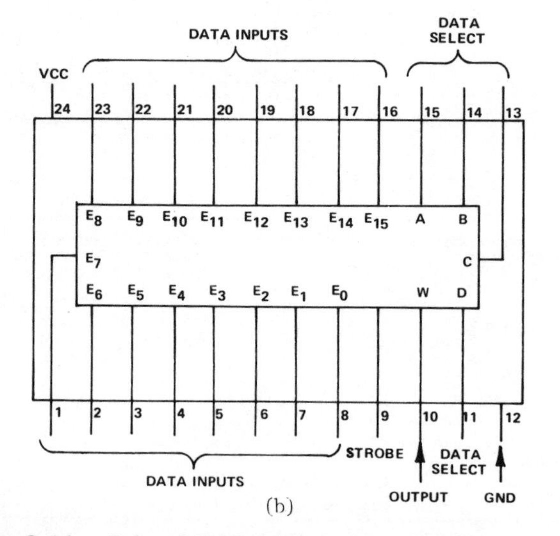

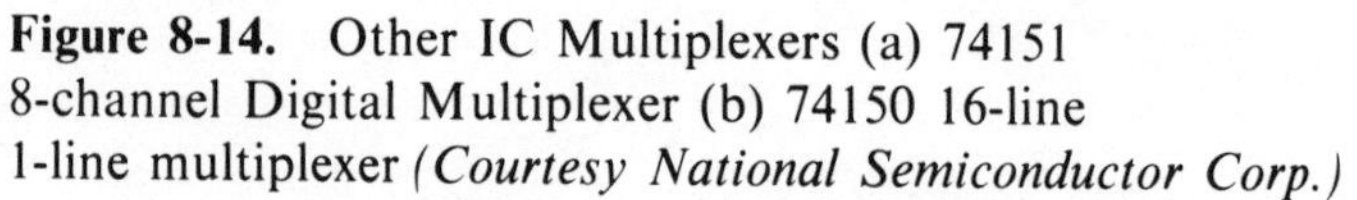

Figure 8-14. Other IC Multiplexers (a) 74151 8-channel Digital Multiplexer (b) 74150 16-line 1-line multiplexer *(Courtesy National Semiconductor Corp.)*

directly following the data present on the selected data input, while output W is complemented. The 74150 IC is a 16:1 multiplexer with an output that is inverted from that of the input. This multiplexer has four data select inputs to be able to select one of the sixteen inputs and is a 24-pin IC. Both of these ICs have a strobe input to enable the chip.

8-6 DEMULTIPLEXERS

A basic demultiplexer is opposite to a multiplexer in that it is a switching device that can select one of a number of outputs and connect it to a single common input as shown in Figure 8-15.

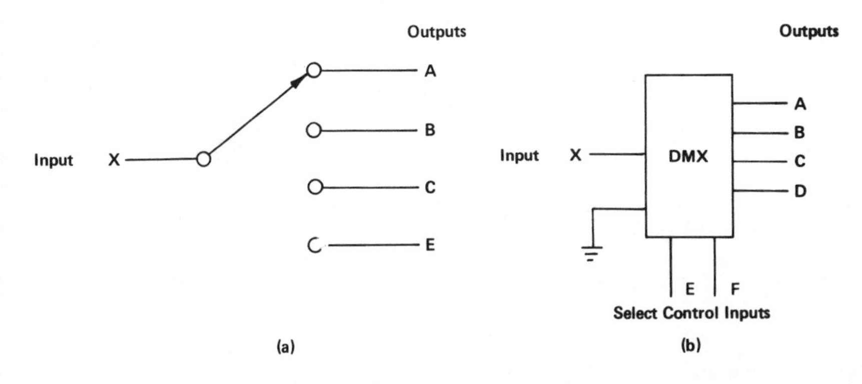

Figure 8-15. Demultiplexers (a) Single-pole 4-position Switch (b) Digital Demultiplexer

The mechanical switch can serve as a multiplexer or demultiplexer, depending on how it is wired. The digital demultiplexer, being a logic circuit, will only allow the data to pass from input to output and, therefore, is a separate device as shown in Figure 8-16.

Similar to the multiplexer, data at input X of the demultiplexer can be routed to any one output A-D by the binary code placed at inputs E and F. The output select truth table shows the code needed to select each output.

The counterpart to the 74153 dual 4:1 multiplexer IC is the 74155 dual 2:4 demultiplexer IC shown in Figure 8-17.

Acting as a double-pole 4-throw switch, a single output of each section of the 74155 is selected with common select inputs A and B resulting in a 2:4 demultiplexer. A single 1:4 demultiplexer can be obtained by permanently wiring one of the strobe inputs high. Data applied to input 1C is inverted at outputs 1Y0-1Y3, while data applied to input 2C is not inverted at outputs 2Y0-2Y3. A 3-line to 8-line decoder or a 1-line to 8-line demultiplexer can be made from the 74155 with inputs 1C and 2C connected together and inputs 1G and 2G connected together.

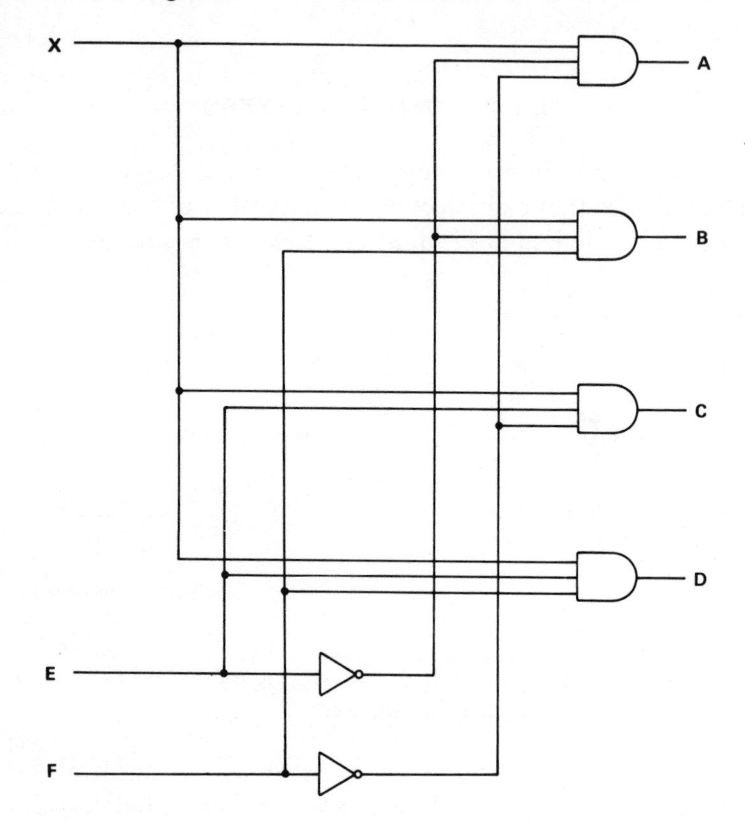

(a)

Desired Output	E	F
A	0	0
B	0	1
C	1	0
D	1	1

(b)

Figure 8-16. Basic Demultiplexer
(a) Logic Diagram (b) Output Select Truth Table

Another popular demultiplexer is the 74154 4-line to 16-line decoder/demultiplexer shown in Figure 8-18. The 74154 IC is the counterpart to the 74150 multiplexer IC and also has 24 pin connections. Inputs A-D select 1 of 16 outputs with the selected output going low, if inputs G1 and G2 are also low. One of these inputs can serve as the strobe, while the other is the data input.

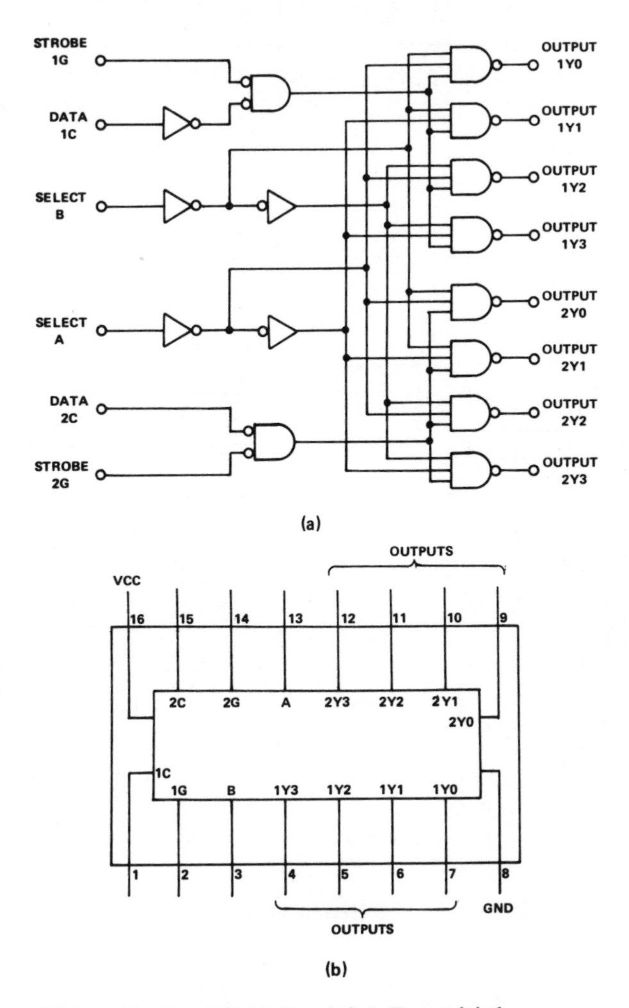

Figure 8-17. 74155 Dual 2:4 Demultiplexer (a) Logic Diagram (b) Pin Configuration (c) 2:4 Line Decoder or 1:4 Line Demultiplexer Truth Table (d) 3:1 Line Decoder or 1:8 Line Demultiplexer Truth Table *(Courtesy National Semiconductor Corporation)*

INPUTS				OUTPUTS			
SELECT		STROBE	DATA				
B	A	1G	1C	1Y0	1Y1	1Y2	1Y3
X	X	H	X	H	H	H	H
L	L	L	H	L	H	H	H
L	H	L	H	H	L	H	H
H	L	L	H	H	H	L	H
H	H	L	H	H	H	H	L
X	X	X	L	H	H	H	H

INPUTS				OUTPUTS			
SELECT		STROBE	DATA				
B	A	2G	2C	2Y0	2Y1	2Y2	2Y3
X	X	H	X	H	H	H	H
L	L	L	L	L	H	H	H
L	H	L	L	H	L	H	H
H	L	L	L	H	H	L	H
H	H	L	L	H	H	H	L
X	X	X	H	H	H	H	H

(c)

INPUTS				OUTPUTS							
SELECT			STROBE OR DATA	(0)	(1)	(2)	(3)	(4)	(5)	(6)	(7)
C†	B	A	G‡	2Y0	2Y1	2Y2	2Y3	1Y0	1Y1	1Y2	1Y3
X	X	X	H	H	H	H	H	H	H	H	H
L	L	L	L	L	H	H	H	H	H	H	H
L	L	H	L	H	L	H	H	H	H	H	H
L	H	L	L	H	H	L	H	H	H	H	H
L	H	H	L	H	H	H	L	H	H	H	H
H	L	L	L	H	H	H	H	L	H	H	H
H	L	H	L	H	H	H	H	H	L	H	H
H	H	L	L	H	H	H	H	H	H	L	H
H	H	H	L	H	H	H	H	H	H	H	L

†C = inputs 1C and 2C connected together
‡G = inputs 1G and 2G connected together

(d)

Figure 8-17. *(Continued)*

SUMMARY

Digital conversion devices use encoders for getting data into a digital system while decoders are used within the system for executing operations and converting the binary data back into useful information.

Multiplexers and demultiplexers used in digital systems help reduce wiring requirements between the various digital units by selecting one of several inputs, transmitting the data over common

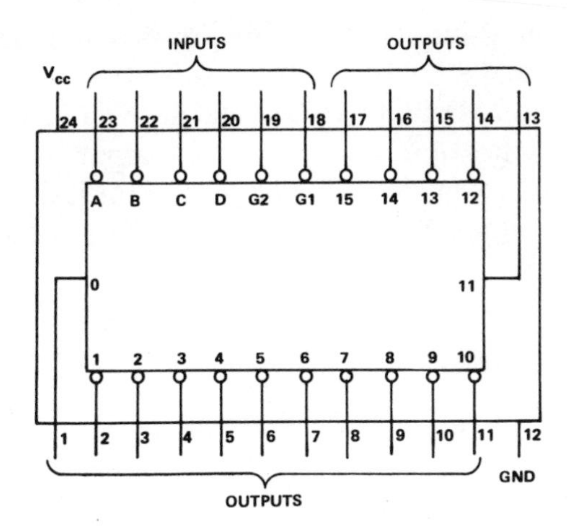

Figure 8-18. 74154 4-line to 16-line Decoder/Demultiplexer *(Courtesy National Semiconductor Corporation)*

lines and selecting the proper one of several outputs, insuring that the data arrives at its destination.

8-7 PROJECT # 8: WIRING AND TESTING DIGITAL CONVERSION AND SWITCHING DEVICES

You will increase your skills in digital circuits greatly if you build and test all of the basic circuits given in this chapter. The circuits given in this project are intended to show possible ways to implement logic circuits and introduce you to the new ICs.

8-7a Testing the NAND Gate Encoder

The NAND gate encoder shown in Figure 8-19 uses a 7432 quad 2-input OR gate to expand the input capability of the 7410 and 7400 ICs used. Refer to Figure 8-2 for the logic diagram of the decimal-to-binary NAND gate encoder.

Input switches A-I are the decimal numbers 1-9 respectively and are normally high. When any one switch is brought low, the binary equivalent will be indicated by the LEDs.

8-7b Testing Decoders

Using two 7476 dual JK Master/Slave flip-flop ICs and a 7408 quad 2-input AND gate wire up and test the circuit shown in Figure

8-4. Combine the 2-input AND gates to produce two separate 3-input AND gates.

The 7442 and 7441 BCD-to-Decimal decoder IC can be tested by using input switches A-D as the B-C-D input and checking the proper output with a single LED and lead or the loco probe. Refer to Figure 8-6 and Figure 8-7 respectively for the pin configuration of these ICs.

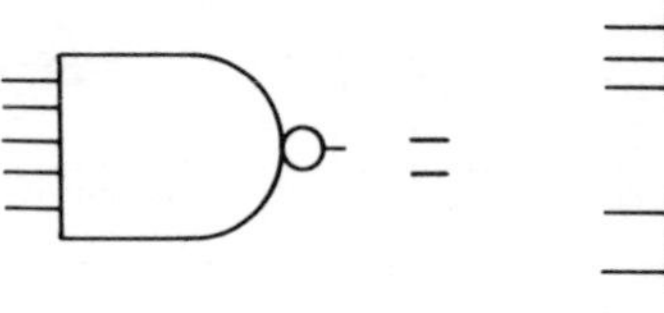

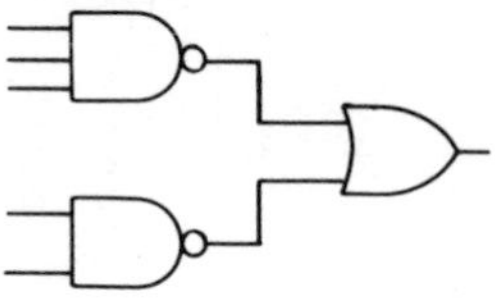

(a)

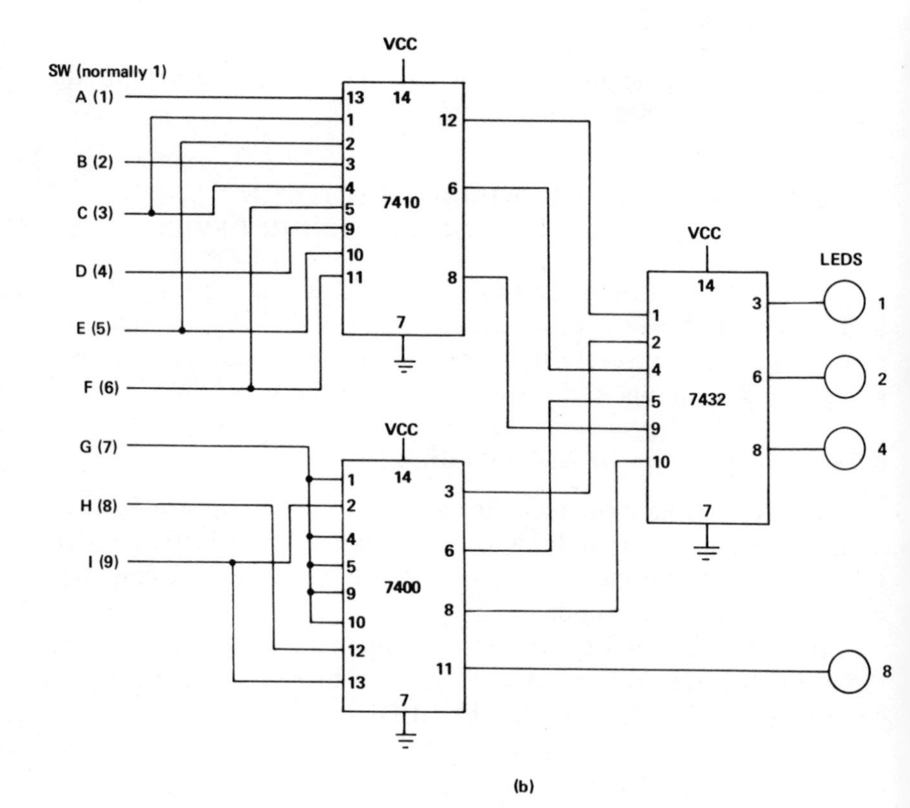

(b)

Figure 8-19. NAND Gate Encoder (a) 5 input NAND gate equivalent circuit (b) Wiring Diagram

Wire up and test each circuit, using Figure 8-8 (pin configuration) and Figure 8-9 (truth table) for the 7446, 7447 and 7448 BCD to 7-segment decoder/driver ICs. These decoders will be used in project # 9 for constructing a binary counter with display.

8-7c Testing Multiplexers

With the knowledge and skills you have acquired construct and test the circuits shown in Figures 8-11 and 8-12.

The 74153 dual 4:1 multiplexer IC can be tested as shown in Figure 8-20.

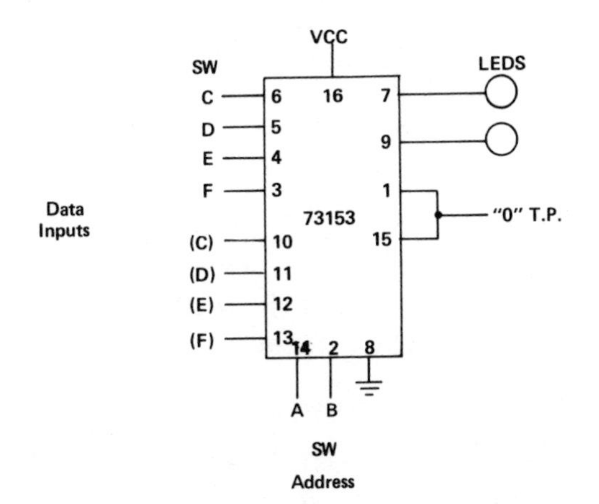

Figure 8-20. Testing the 74153 Multiplexer

Referring to Figure 8-13 and the truth table, use the following procedure to test each section of the 74153 IC.

1. Select the desired address with switches A and B
2. Select and place in the high position respective input switch C-F.
3. Press the "0" T.P. switch and the LED connected to the respective output should light.
4. Repeat steps 1-3 for each address.
5. Test the other section of the IC by shifting input switches C-F to pins 10-13 respectively and repeat steps 1-4 above.

8-7d Testing Demultiplexers

Construct the basic demultiplexer shown in Figure 8-16 and prove the output select truth table.

The 74155 dual 2:4 demultiplexer IC can be tested as shown in Figure 8-21.

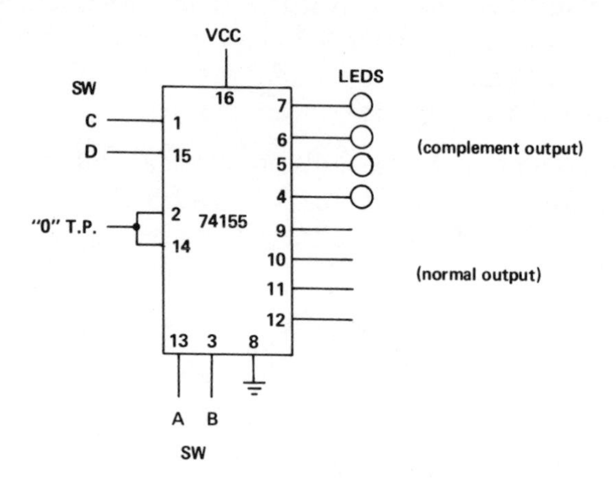

Figure 8-21. Testing the 74155 Demultiplexer

Referring to Figure 8-17 and the truth tables use the following procedure to test each section of the 74155 IC.

1. Select the desired address with switches A and B.
2. Press and hold the "0" T.P. switch.
3. Operate proper input switch, i.e.;
   For section 1, operate input switch C and the LED connected to the selected output will extinguish when the input is a 1.
   For Section 2, operate input switch D and the LED connected to the selected output will go on when the input is a 1.
4. Repeat steps 1-3 for each address.
5. Test the other section of the IC by shifting output LEDs to pins 9-12 respectively and repeat steps 1-4 above.

8-7e Testing a Basic Data Transmission System

The basic data transmission system shown in Figure 8-22 is an example of how multiplexers and demultiplexers are used to reduce wiring requirements. This system uses a 4:1:4 ratio, whereas, as the ratio increases, say 32:1:32, the system becomes more efficient.

Two 7476 dual JK flip-flop ICs are wired to produce MOD-4 counters which are connected to the address inputs of the 74153 multiplexer and the 74155 demultiplexer. The clock synchronizes each counter so that the selected input line to the multiplexer matches the selected output line of the demultiplexer. These lines are constantly being scanned, looking for data. The outputs of the 74155 IC are normally high and it is easier to see a low at one of the inputs of the 74153 IC. Therefore, all of the input switches are normally high.

To test the system, set the clock to a low frequency and move one of the input switches to a low. The appropriate LED should blink as its channel is scanned and the data present is a 0. If all of the input switches are brought low, the LEDs will show a ripple effect similar to a ring counter.

If you desire to increase the system to a 8:1:8 ratio, inverters will be needed at section 1 outputs (pin 4-7) of the 74155 IC to make all outputs of the system consistent.

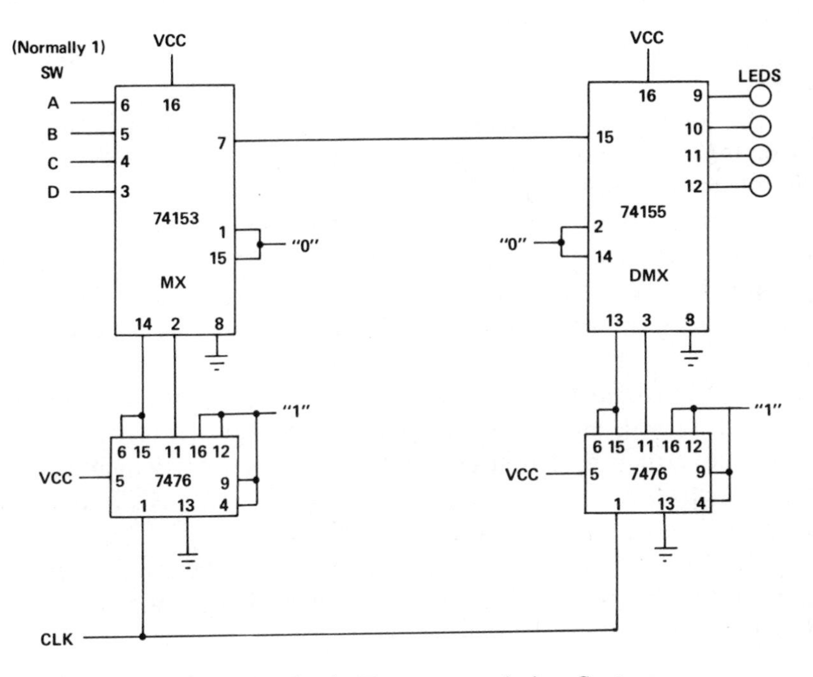

Figure 8-22. Basic Data transmission System

CHAPTER 9

Obtaining Information from Digital Systems Using Optoelectronic Displays

Optoelectronic devices combine the technologies of optics and electronics and have become one of the fastest-growing segments of the electronics industry.

Initially, photodetectors were developed, such as photoelectric cells (producing an electric potential when exposed to light), photovaristors (changing resistance when exposed to light) and photodoides and phototransistors (emitting current when exposed to light).

It was found that germanium and silicon diodes, which were forward biased, emitted small portions of light. Improvements in diode construction produced light-emitting diodes which you will learn to use in this chapter. Also included are gas discharge indicators, fluorescent and liquid crystal displays.

In Project # 9 you will learn how to test optoelectronic displays and build a binary counter with decimal readout, using some of the ICs studied in previous chapters.

9-1 LIGHT-EMITTING DIODE (LED)

Light-emitting diodes (LEDs) are mostly made of Gallium Arsenide, (GaAs) and Gallium Arsenide Phosphide, (GaAsP) material which emits more light than regular diodes. The phenomenon of light emitted from a PN junction occurs when an electron crossing the junction recombines with a hole. The energy that is given up when the electron falls to the lower energy level results in a photon of light. Of course, many electrons must be recombining in order for light to be visible with the naked eye. The schematic diagram for an LED is shown in Figure 9-1a. Light will be emitted only when the LED is forward biased.

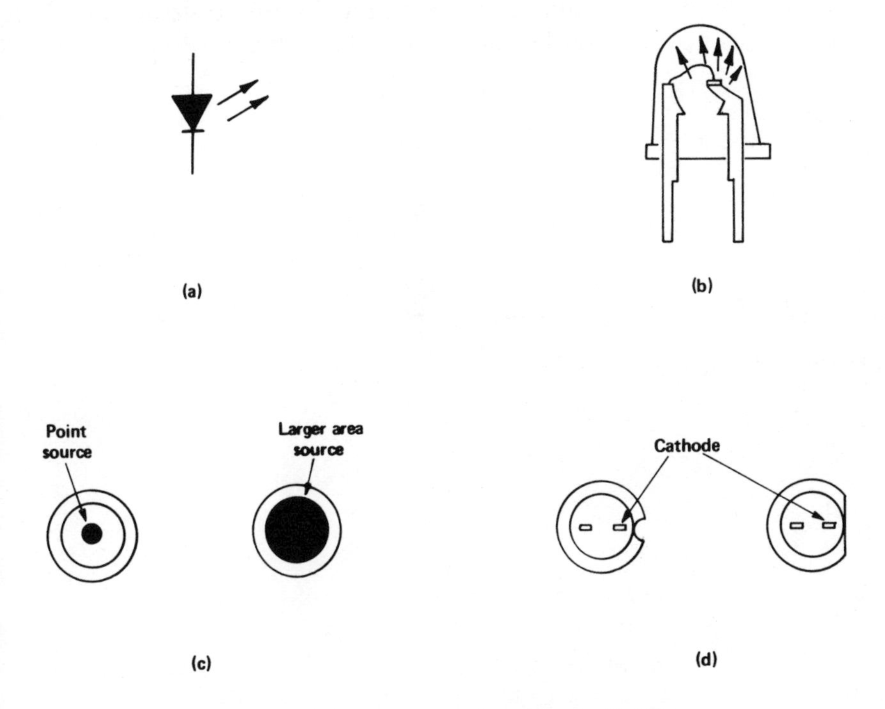

Figure 9-1. Light-emitting Diode (LED)
(a) Schematic Diagram (b) Physical View
(c) Clear Lens vs. Diffused Lens
(d) Index for Cathode

The colors of LEDs commonly found are red, orange, yellow and green. Blue LEDs are available, but are usually more expensive.

The LED is constructed on the top of one lead with a fine wire connected to the other lead and the entire assembly is encapsulated in a clear epoxy plastic as shown in Figure 9-1b. The epoxy forms a lens which magnifies the emitted light. This lens is often diffused to give a larger area source of light as shown in Figure 9-1c. The cathode of an LED is mostly denoted by an index notch or flat side on the epoxy as shown in Figure 9-1d.

It is necessary to limit the amount of current through an LED with a limiting resistor as shown in Figure 9-2a. Generally, the continuous forward current in LEDs is from 5 ma to 40 ma, and the forward voltage (V_F) drop of LEDs ranges from 1.65 to 2.2 volts. If you decide to use a LED with a 5-volt power source, such as the pilot light for the breadboard/tester given in Project # 1, you could use the formula given to find the value of the limiting resistor. Say that V_F = 1.8 volts and you let the current through the LED be 15 ma,

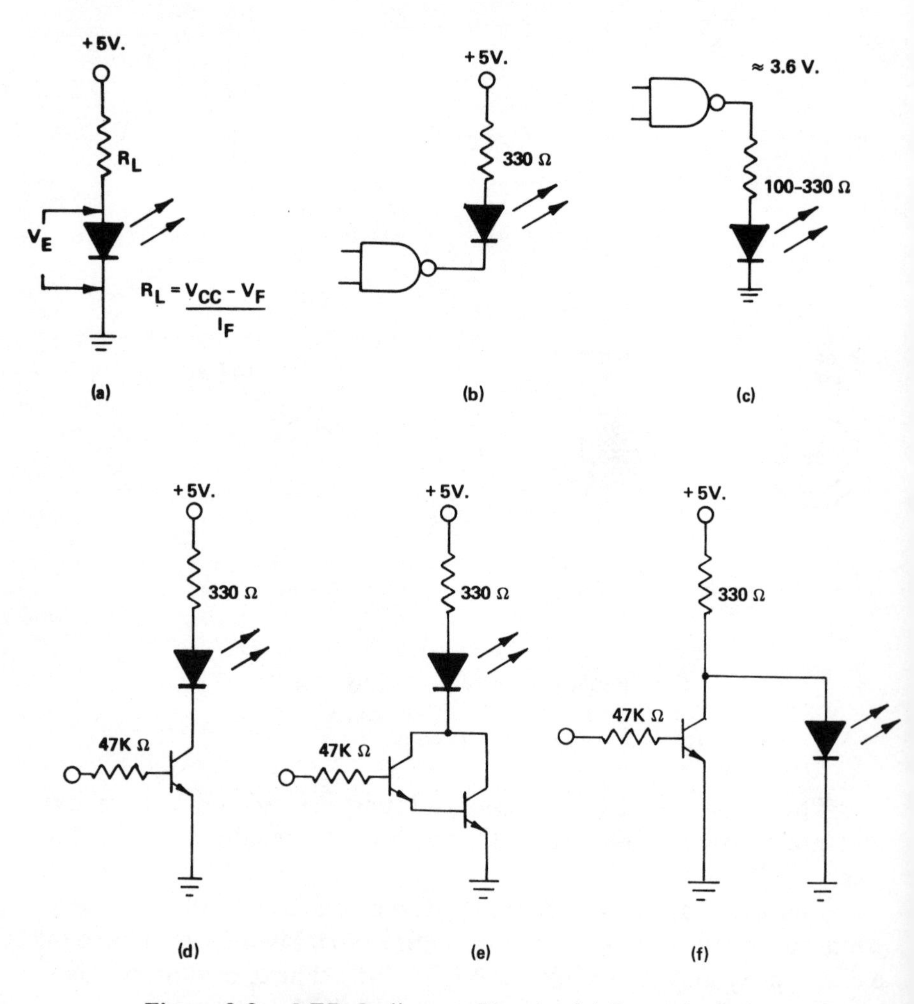

Figure 9-2. LED Indicator Circuits (a) R_L calculation (b) Low Indication (c) High Indication (d) With Transistor Driver (e) With Darlington Pair (f) Low **"Shunt"** Indicator

then the value of R_L would be 213 ohms. A 220-ohm resistor could be more easily obtained and not affect the circuit operation appreciably This type of LED indicator can be used directly from a TTL IC when no other loads are attached as shown in Figure 9-2b. A low output will cause the LED to light and the current is less than 10 ma to protect the IC. An alternate method of indicating a high output is shown in Figure 9-2c. The limiting resistor can range from 100 to 330 ohms for satisfactory operation.

To reduce the loading effect on an IC, you can use a transistor driver or a Darlington transistor pair driver as shown in Figures 9-2d and 9-2e respectively. With these arrangements, less than 1 ma of current is drawn from the IC, while the LED still has the normal amount of current in order to light. A low output indicator driver is shown in Figure 9-2f. When the transistor is off, current flows through the LED causing it to light. When the transistor is on, the current is shunted away from the LED causing it to extinguish.

There are various methods of mounting LEDs as illustrated in Figure 9-3. A black grommet enhances the visibility of the LED, but the particular mounting technique is left to your discretion and need.

9-2 LED SEVEN-SEGMENT DISPLAY

The popular seven-segment display used with digital clocks, watches, calculators, digital voltmeters, avionics and computers and other digital readout displays are compatible with integrated ciruits.

Two styles of LED seven-segment displays are shown in Figure 9-4. Discrete LEDs are arranged to form the seven segments shown in Figure 9-4a. When a particular segment is selected, all four LEDs will light. The other style shown in Figure 9-4b uses solid diffused reflective bars which are connected with light pipes to individual LEDs. A decimal point, or a plus and minus sign, is often added to the display which is contained on a 14-pin DIP IC, but all of the pins are usually not used.

The LED seven-segment displays are of two types, the common anode and the common cathode, which are shown schematically in Figures 9-4c and 9-4d respectively. The common anode type of display requires lows to turn on the segments, while the common cathode requires highs to turn on the segments. These LED displays also need external limiting resistors as shown.

The various segments needed to produce the ten decimal digits are shown in Figure 9-4e. The 7446, 7447 and 7448 BCD-to-7 segment decoder/drivers ICs studied in the last chapter are used to display the decimal digits. Nine distinct letters are also possible with this display.

The LED seven-segment displays generally are red, green, orange or yellow with character sizes ranging from .11-1.0 inch. Black plastic caps are sometimes placed over the displays to provide better contrast as shown in Figure 9-5.

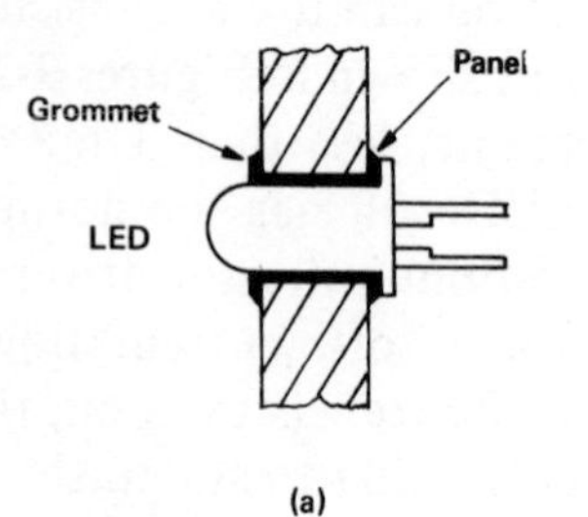

(a)

Epoxy
Panel
LED

(b)

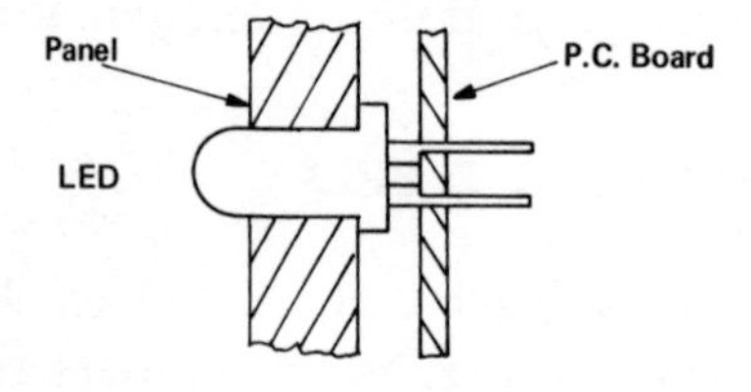

(c)

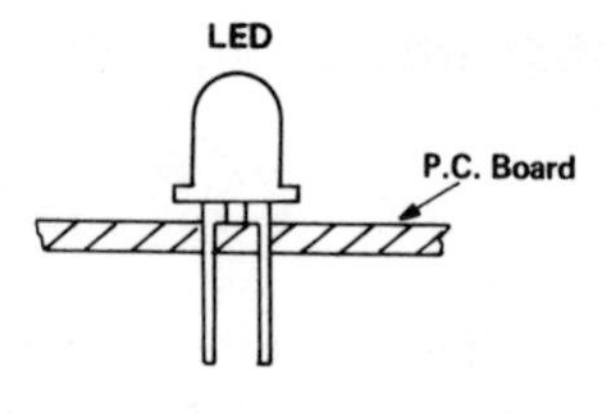

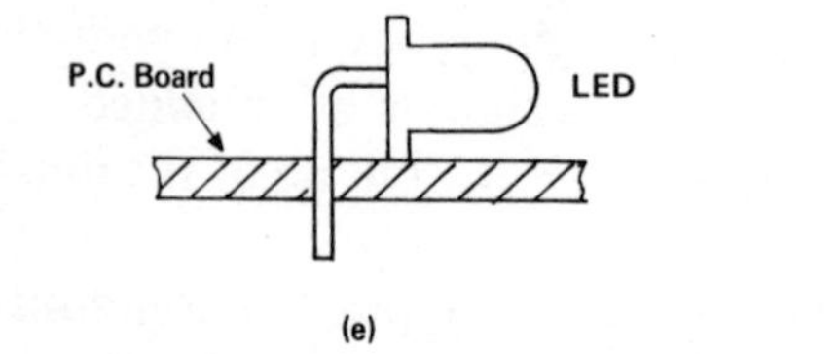

(e)

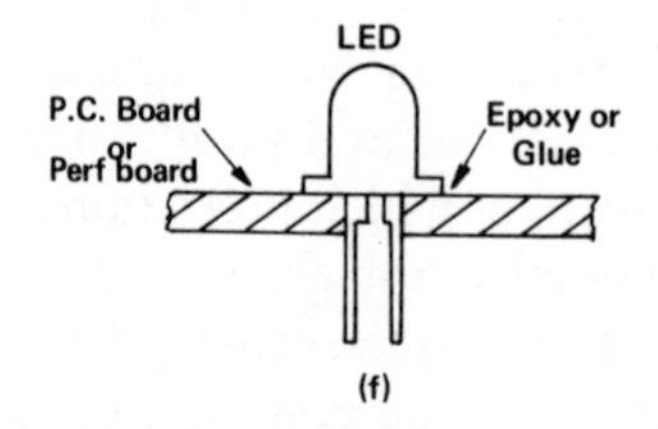

(f)

Figure 9-3. Methods of mounting LEDs (a) With Grommet (b) With Epoxy (c) With P.C. Board (d) Mounted Upright (e) Mounted with Bent Leads (f) Flush Mounted

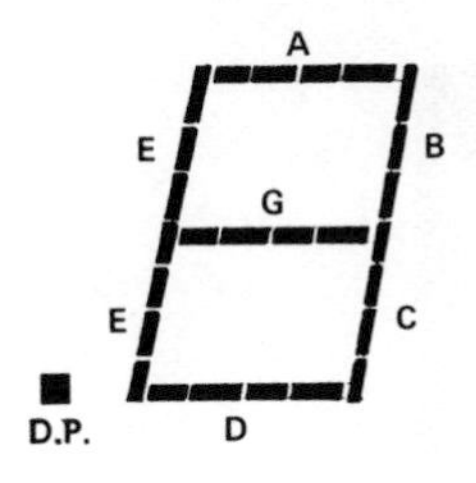

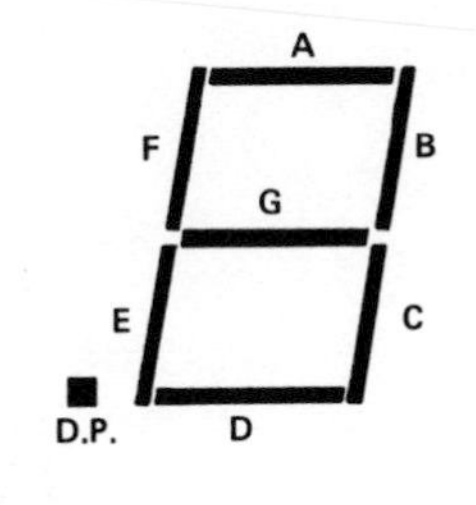

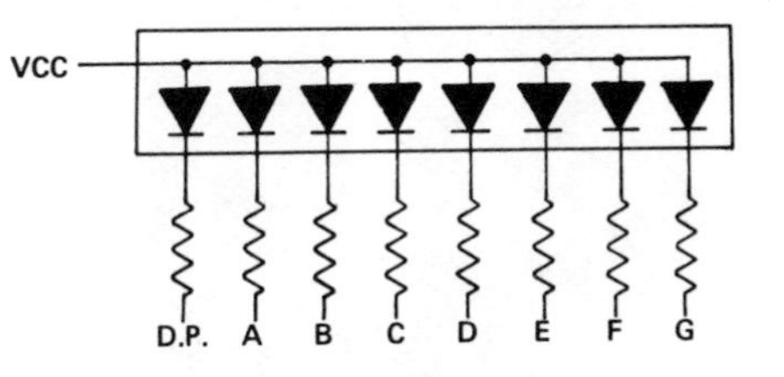

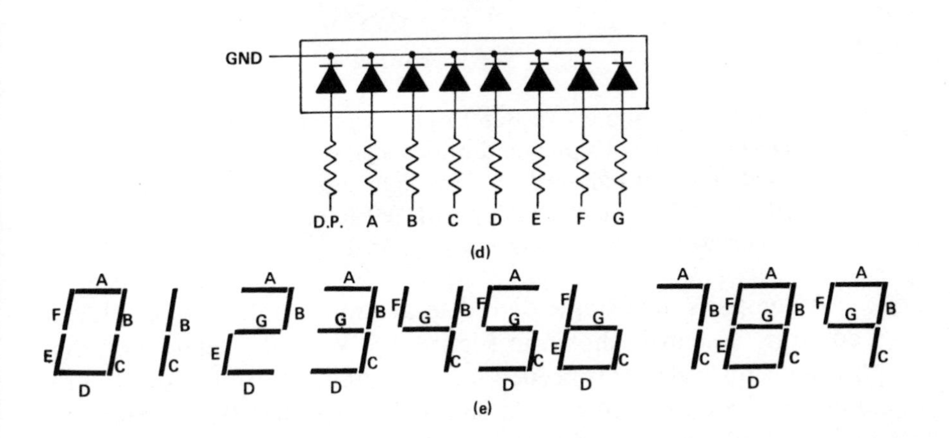

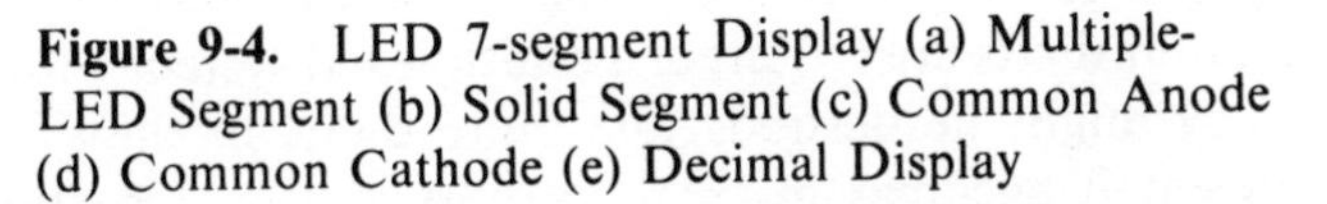

Figure 9-4. LED 7-segment Display (a) Multiple-LED Segment (b) Solid Segment (c) Common Anode (d) Common Cathode (e) Decimal Display

Multi-digit LED displays, as shown in Figure 9-6a, reduce power consumption and cost. Each digit has a built-in "bubble" magnifier that directs light to the eye. The displays are equipped for multiplex operation where the seven anode segments are common to all digits and a specific digit can be turned on by selecting the proper cathode.

Figure 9-5. Reflective Bar LED 7-segment Display (top): Decimal one with plus and minus sign.
(middle:) Cap removed to show light pipes.
(Bottom) 7-segment Display with left-hand decimal point.
(Courtesy Litronix Incorporated)

A comparison of single-digit displays and a multi-digit display in a counting system is shown in Figure 9-7. With the single-digit display, a decoder/driver is needed for each digit. The input is to the 1's counter. The output of the 1's counter drives the 10's counter, the output of 10's counter drives the 100's counter, and so on. When more than four digits are required, a multiplexed system with a multi-digit display might be desired. The more digits in the system, the more efficient it becomes in terms of cost and construction. With multiplexing, as shown in Figure 9-7b, each digit in the display shares the same decoder/driver. The main consideration is that the address input to the multiplexer and the digit sequences be synchronized so that the output from a counter coincides with its respective digit that is being turned on by the sequencer. The scan rate of the clock is from 100 Hz upward and the segments are "refreshed" (relighted) about one hundred times a second so that they give the impression they are permanently lit.

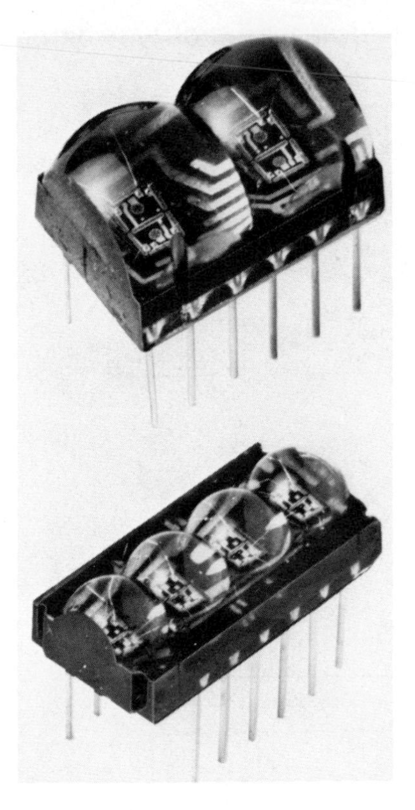

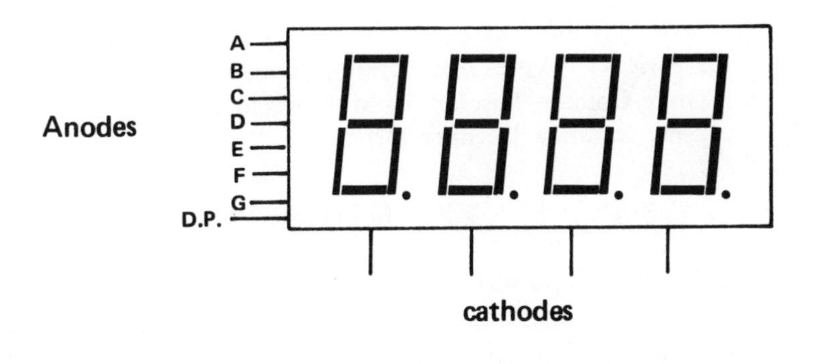

(b)

Figure 9-6. Multi-digit LED Displays (a) 2-and 4-digit LED Displays *(Courtesy Litronix Incorporated)* (b) Schematic Representation

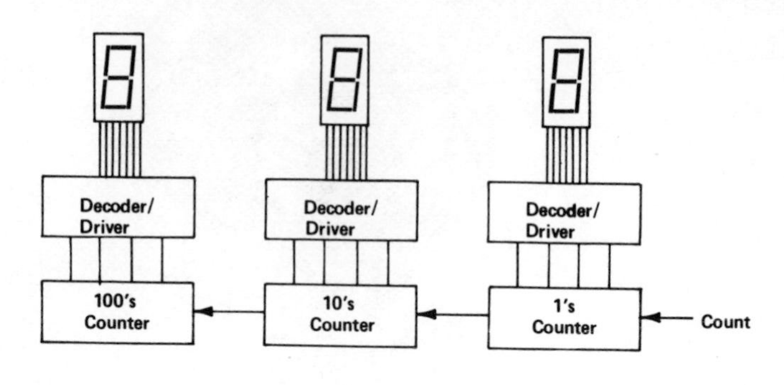

(A)

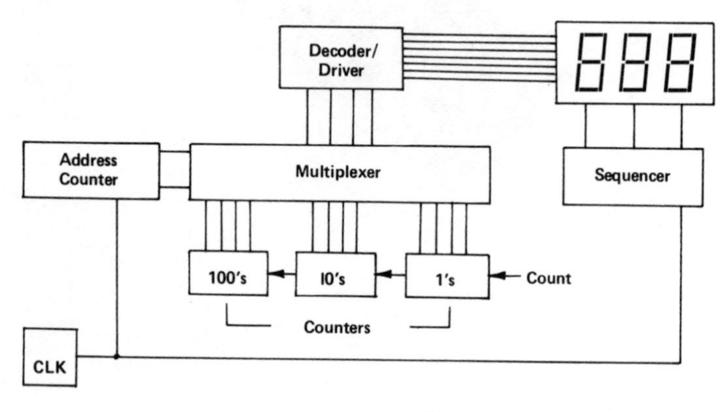

(B)

Figure 9-7. Multi-digit Display systems (a) Direct Drive Counter Display (b) Multiplexed Counters using common Decoder/Driver

9-3 LED 5 X 7 MATRIX DISPLAY

A LED display capable of producing numbers, letters and special characters is the Monsanto MAN-2A, red alpha-numeric display shown in Figure 9-8. The LED's are arranged in a 5 × 7 matrix where the rows are the cathodes and the columns are the anodes. To light the upper left-hand LED, row 1 (pin 2) would have to be brought low and column 1 (pin 5) would have to be brought high. In order to produce a character, the various rows are brought low, while each column is brought high with a scanning operation. The letter F is shown in Figure 9-9a. When column 1 is high, all 7 rows are low, when column 2 is high, rows 1 and 4 are low, when column 3 is high, rows 1 and 4 are low, when column 4 is high, rows 1 and 4 are low, when column 4 is high, rows 1 and 4 are low, and when column 5 is high, only row 1 is low. A RAM or ROM is usually connected to the 7 row inputs to produce the desired character as the columns are scanned. Project #9 includes such a system for experimentation. Some other characters are also shown in Figure 9-9b.

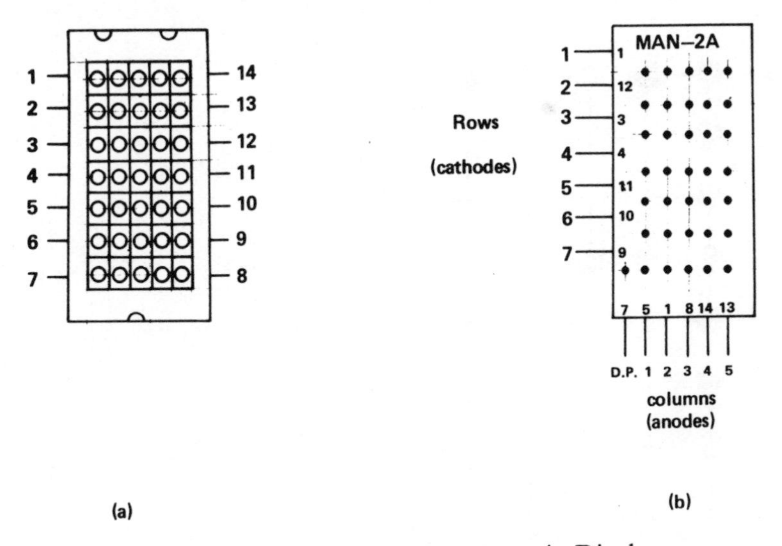

Figure 9-8. MAN-2A Alpha-Numeric Display
(a) 14-pin DIP (b) Schematic Representation

Each LED of the MAN-2A is rated at 20 ma maximum current, so it will be necessary to include limiting resistors in the anode lines. A 330-ohm resistor limits the current through a single LED to about 10 ma, using +5V as V_{cc}. As other LEDs are turned on, the total current through the resistor varies little. The nearly constant forward voltage drop of the LEDs causes the initial total current to divide between the LEDs that are turned on. The brightness of each LED is reduced as more LEDs are turned on in a single column. If a TTL IC is used to drive the MAN-2A, the value of the limiting resistors may have to be reduced to insure the desired brightness.

9-4 GAS DISCHARGE DISPLAYS

The gas discharge display operates on the principle of the cold cathode neon glow indicator. The numerical indicator shown in

Figure 9-10 consists of a grid-type anode and ten separately formed metal cathodes in the shape of the numerals 0-9, which are stacked one behind the other. Decimal points are sometimes included as shown. This entire array is placed in a vacuum tube which contains neon gas.

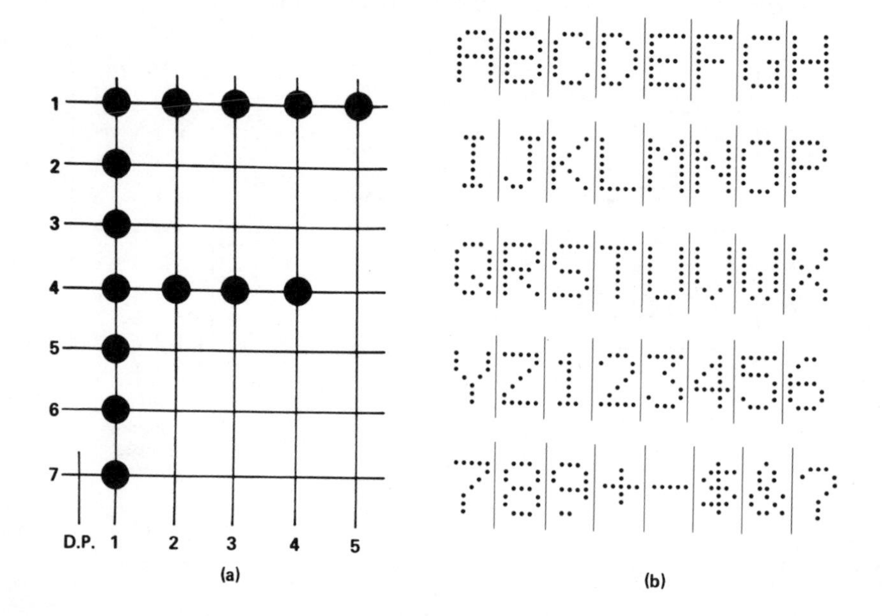

Figure 9-9. 5×7 LED matrix (a) Method of Selection (b) Some Characters

A positive voltage of approximately 150 to 200 volts is placed on the anode of the tube via the current limiting resistor R_L. When a cathode is grounded or brought low, a neon gas discharge (glow) appears near the surface of that cathode, thus displaying the corresponding numeral. Transistor drivers or special integrated circuits such as the 7441, are needed to energize the cathodes.

Gas discharge displays are also available in alphabetical characters, special characters, seven-segment and multi-digit panels.

The major advantage of gas discharge displays is that they can be read from a relatively far distance and in fairly high ambient light conditions. However, gas discharge displays require high voltages, such as normal vacuum tubes, and can only be read with the least confusion by viewing them from a head-on position since the glowing numeral is viewed through the front character cathode wires.

(a)

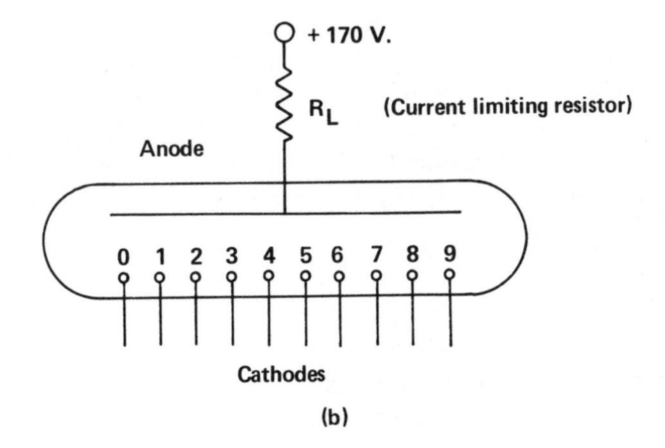

(b)

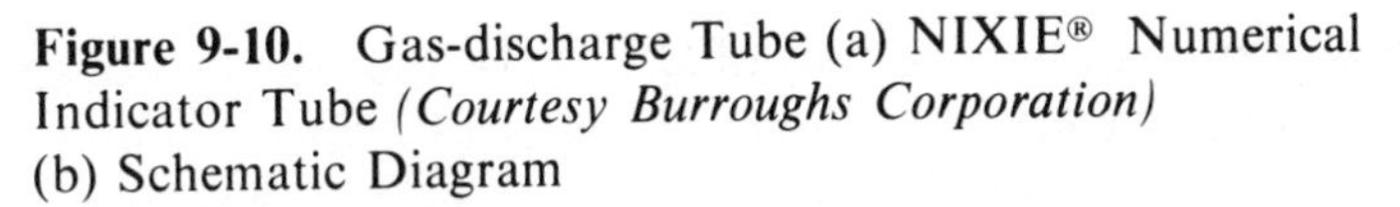

Figure 9-10. Gas-discharge Tube (a) NIXIE® Numerical Indicator Tube *(Courtesy Burroughs Corporation)* (b) Schematic Diagram

As with other electronic components, gas discharge displays should be wired or constructed according to the manufacturer's specification sheet and kept within the maximum allowable voltage and current requirements.

9-5 OTHER TYPES OF DISPLAYS

Seven-Segment Incandescent-Lamp Display

The seven-segment incandescent-lamp display, such as the RCA Numitron, uses short filament wires supported against a dark background and is encased in a 9-pin miniature vacuum tube. The operating voltage is between 3.5 and 5 volts and requires about 24 ma per segment. The segments glow white and the segment voltage may be varied to control brightness. The segment can be driven by transistors or a low output decoder/driver capable of handling the required current. A decimal point is also formed from short filaments.

Fluorescent Display

Fluorescent displays are made of the seven-segment and multidigit types of displays. The seven-segments anodes are made of a fluorescent material and mounted on an insulated background. Two directly heated cathode wires (filaments), or a cathode grid, are placed in front of the seven-segment pattern and the entire unit is encased in a miniature vacuum tube. When the proper segments are selected, electrons flow from the filaments to those segments which, in turn, give off a greenish glow indicating the desired numeral or character. The operating voltages and currents vary depending on the size of the display. A typical display, .57 inches in height, would require 1.5 volts for the filaments, 22 volts for the segments and draw .5 ma of current per segment. Many fluorescent displays are compatible with TTL 7400 series ICs.

Liquid Crystal Display

The liquid crystal display (LCD or LXD) is of particular importance to digital electronics because it only draws nanoamperes to microamperes of current, therefore, a good reason for their use in digital wristwatches.

Another unique feature of LCD is that it is not a light-generating device as are all the other types of displays. The LCD depends on good ambient light or some form of other non-ambient light source. Unlike the light-generating displays that tend to "wash out" as the ambient light increases, the LCD becomes increasingly more legible.

There are two types of liquid crystal displays. The first was the dynamic-scatter LCD recognized by its milky white characters. The newer, and more often used, field-effect LCD has either black or clear characters.

The basic construction and operation of a field-effect LCD is shown in Figure 9-11. A nematic liquid (liquid crystal) is placed between two panels of glass which have a microscopically thin layer of metal deposited on their facing sides. This metallization is so thin that the glass still appears transparent. The metallization completely covers the backplane panel of glass, while it is in the form of seven-segment displays, colons and perhaps alpha-numeric characters on the other panel of glass. With the field-effect LCD two polarizing filters are placed on each side of the glass to cause the light to pass through the unit on a direct path. With a transmissive system, light

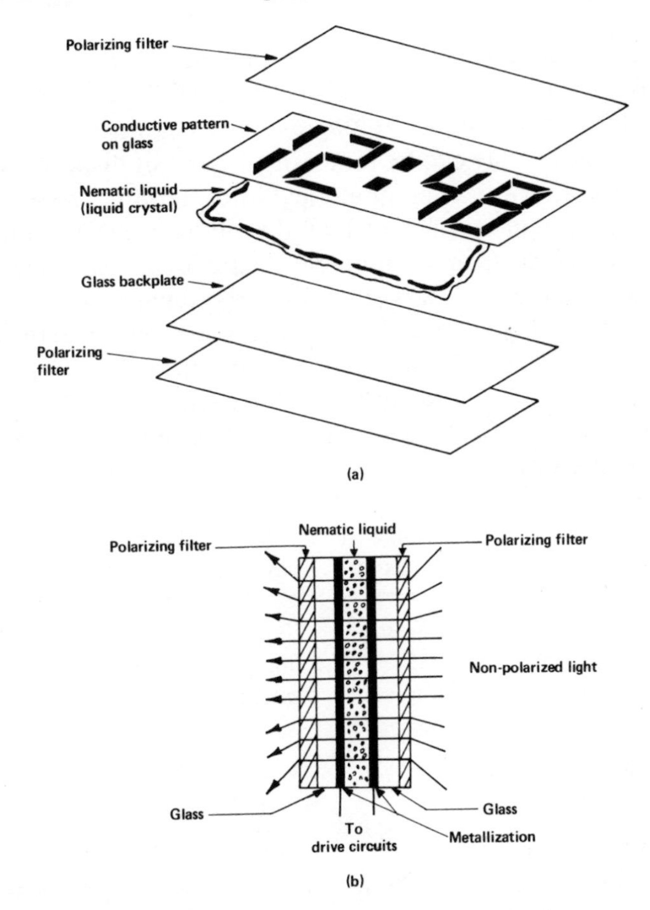

Figure 9-11. Field-effect Liquid Crystal Display
(a) LCD Construction (b) Transmissive LCD Operation

passes right through the display itself, whereas sometimes a reflector is placed behind the unit to produce a reflective system where the light is reflected back through its front panel. The entire unit is sealed and has edge-connectors for the electrical connections.

When the desired segments are energized with AC or a train of square wave pulses, the nematic liquid beneath these segments rotates, or "twists" 90° to alter the light passing through. Light is blocked by this twisting action and the desired character will appear either black or clear.

The main disadvantage of LCDs, is that it takes approximately 5-25 msec for the nematic liquid to polarize and 100-200 msec to depolarize.

SUMMARY

Optoelectronic displays convey the information in a digital system to the outside world. All of the types of displays mentioned in this chapter can be found in single-digit or multi-digit form. The seven-segment display can be found in all types of displays and is the most popular for indicating numerical information. The 5 × 7 LED matrix display can produce all of the alpha-numeric and special characters and its principles of operation are used in cathode ray tube displays associated with larger computer systems.

The LED displays seem to be the most suited with TTL integrated circuits and do not require any special power supplies. However, the actual digital system will determine which type of display is the best to use. The following chart gives a general comparison of the different displays.

Type of Display	Voltage Required	Current Required	Turn ON Time*	Package Material
LED	1.7-5v	20ma	fastest	epoxy
fluorescent	1.5v filaments 22v-anodes	.5ma/ segment	fast	glass
incandescent lamp	3.5-5v	24 ma/ segment	medium	glass
gas discharge	150-200v	15 ma	slow	glass
liquid crystal	3-12vrms	na-2 μ a	slowest	glass

*Turn on-turn off time in relation to the different displays.

9-6 PROJECT #9: TESTING DISPLAYS AND BUILDING A BINARY COUNTER WITH DECIMAL READ-OUT

In order to be able to test optoelectronic displays, you should have the respective manufacturer's specification sheet or be able to ascertain the pin configuration.

Basic Testing of Displays

The basic testing of displays is shown in Figure 9-12. When testing the seven-segment common anode display (Figure 9-12a), connect all anode pins to +5 volts and then with one end of a 330-ohm resistor connected to ground and the other end to a clip lead,

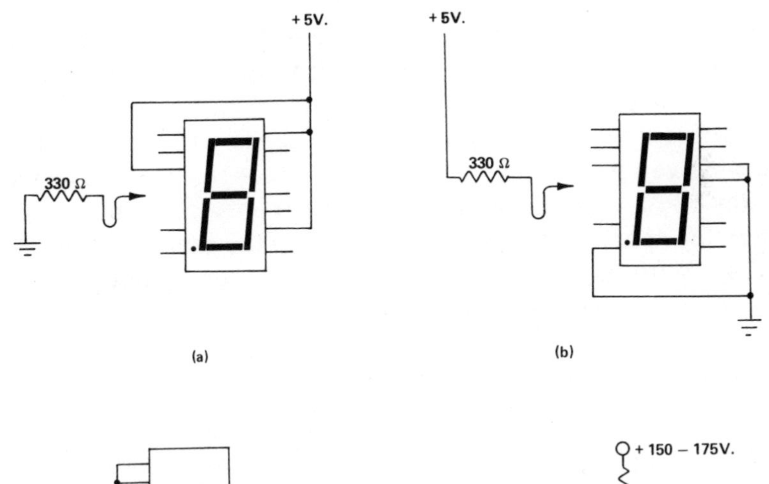

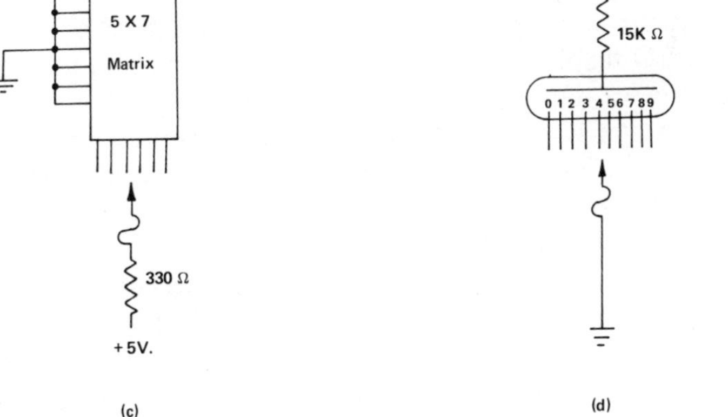

Figure 9-12. Basic Testing of Displays (a) 7-segment common anode LED (b) 7-segment common cathode LED (c) 5x7 matrix LED (d) Gas-discharge Tube

touch all segment pins for a visible indication. With a seven-segment common cathode display (Figure 9-12b), connect all cathode pins to ground and then with one end of a 330-ohm resistor connected to +5 volts and the other end connected to a clip lead, touch all segment pins for a visible indication.

The 5 × 7 matrix LED display (Figure 9-12c) is tested by connecting all of the cathode pins to ground and then with one end of a 330-ohm resister connected to +5 volts and the other end connected to a clip lead, touch all of the anode column pins. All seven LEDs in each column should light.

The gas discharge tube (Figure 9-12d) is tested by connecting the anode to +150-175 volts via a 15K-ohm resistor and with a clip lead connected to ground, touch each cathode to display the desired numeral.

Remember to test the decimal point for each type of display. Multi-digit displays can be tested in a similar manner, but each digit has to be selected and tested separately.

In Figure 9-13, you will see how decoder/drivers ICs are connected to various displays. The main point to remember is the type of display you are using and how it is turned on. Accordingly, you can match the appropriate decoder/driver with the proper output for the display. With multi-digit displays, you will need a multiplex system with a clock, a counter and a sequencer.

Basic Alpha-numeric Display System

A basic alpha-numeric display system is shown in Figure 9-14. Two 7489 RAM ICs are connected together and are simultaneously addressed to provide seven outputs to drive the MAN-2A 5 × 7 matrix LED display. A 7493 counter IC is wired as a mod-5 counter that addresses the RAMs and also drives a sequencer, using a 7441 decoder/driver IC. The 7404 hex inverter IC is used to invert the low outputs of the 7441 IC to drive the anodes of the display. The fifth sequence pulse from the 7441 IC is applied via the 7404 IC back to reset the 7493 counter IC.

After the system is wired, you will have to construct a program truth table for the desired character to be displayed by referring to Figure 9-9. Input switches A-H are used to enter the data into the RAMs. Switch I, which is normally high, is used to store the data. During the time the RAMs are being programmed, the input to the 7493 counter IC (pin 1) is connected to the "0" trigger pulse. This enables each address to be consecutively selected and programmed,

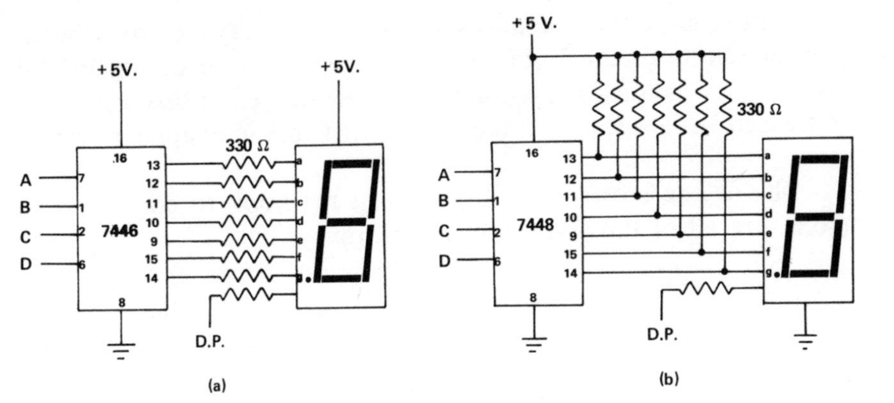

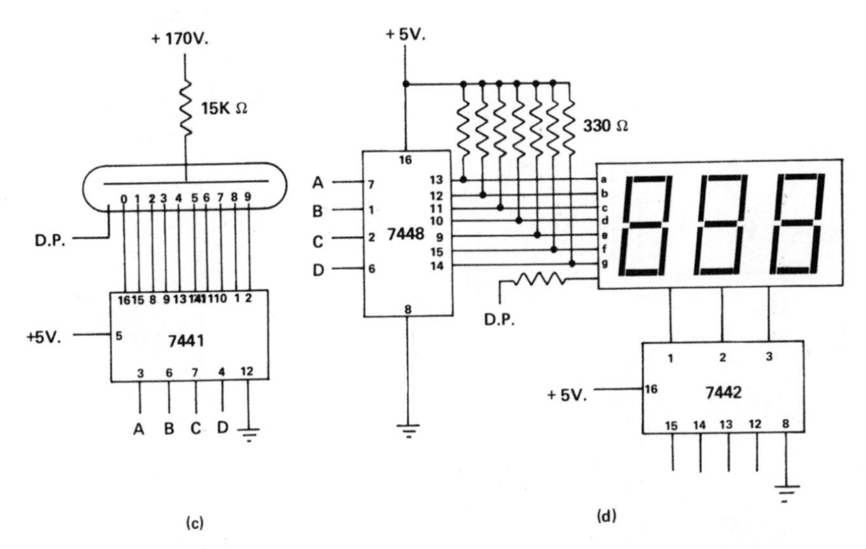

Figure 9-13. Display-Decoder/Driver Applications (a) Common anode 7-segment LED (b) Common cathode 7-segment LED (c) Gas-discharge Tube (d) Multi-digit common cathode LED

while observing the desired output for each column of the display. The LEDs connected to the output of the counter help to identify the address location being programmed.

To display the data stored in the RAMs, the input to the counter is switched to the clock, which is set at about 100hz. As the data from the first memory location is applied to the cathodes of the MAN-2A, the first column is also selected by the sequencer. The LEDs will light in the first column for any corresponding cathode that is low (remember that a stored 1 output of the 7489 RAM is low). The counter then advances to the next memory location, the sequencer selects the next column, and the process occurs again. All five memory locations and columns are sequenced in this order and "refreshed" at a rate that makes the desired character appear permanent.

This is the most difficult system in the book to construct, so make sure all of the connections are firm.

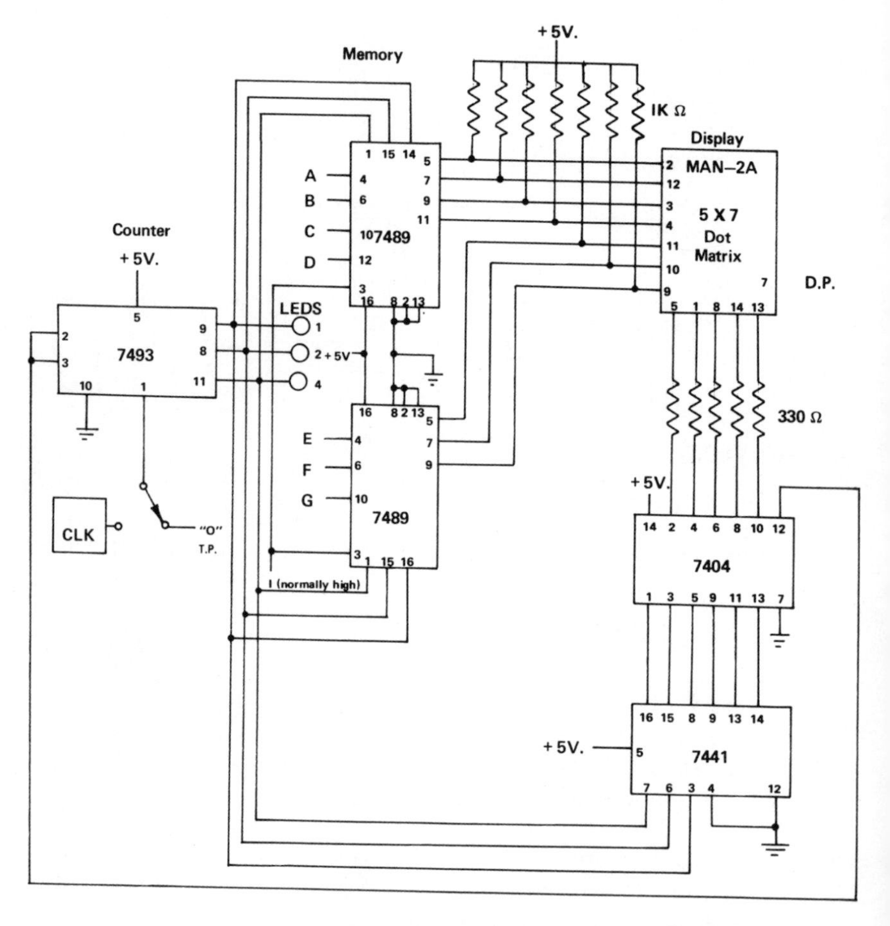

Figure 9-14. Basic Alpha-numeric Display System

Binary Counter with Decimal Read-Out

The binary counter with decimal read-out shown in Figure 9-15 uses a 7490 decade counter IC, a 7447 BCD to 7-segment decoder/driver IC and a MAN-1 LED display. The LED lamps are connected to the outputs of the counter to indicate the binary number are optional. The count (clock) is applied to pin 14 of the counter. Before you begin the count, check all seven segments on the display by bringing SWA (pin 3, lamp test of IC 7447) low. The number eight should be displayed. Be sure to bring SWA high before the counting begins.

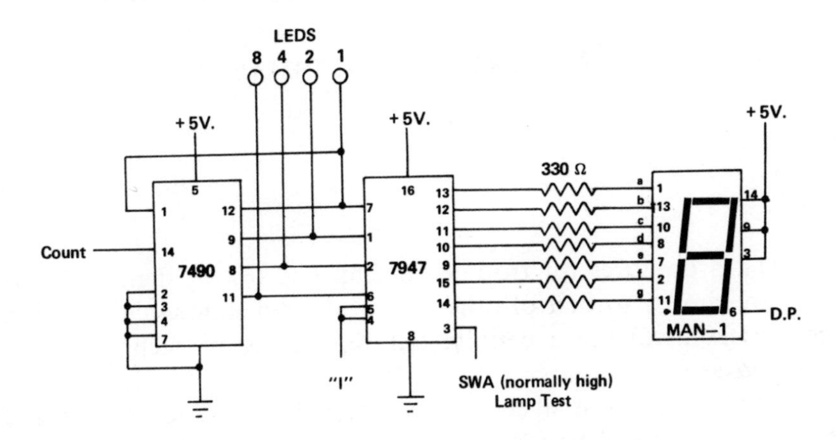

Figure 9-15. Binary Counter with Decimal Read-out

CHAPTER 10

Procedures for Troubleshooting Digital IC Circuits

The procedures for troubleshooting and testing digital IC circuits are similar to that of other types of solid-state circuits and in most cases involving a definite failure, the testing techniques are much easier.

Unlike conventional circuits, you must know what the output of a digital circuit will be for a set of given inputs. With this in mind, you may have to apply different sets of inputs in the form of 0s and 1s before a faulty component is located or completely tested.

Chapter 10 will show some of the common failures that occur in digital ICs and some basic troubleshooting techniques for finding and remedying these malfunctions.

Project 10 shows how to construct two basic digital test instruments, the "LOCO" pulser and the "LOCO" clip, that can be used with the "LOCO" probe to find digital IC failures easier.

10-1 TEST EQUIPMENT USED IN DIGITAL CIRCUITS

Standard electronic test equipment such as an oscilloscope, VOM, VTVM, pulse generator and common hand tools can be used to service digital circuits. Other special digital test instruments, such as the logic probe, logic clip and logic pulser are smaller and generally easier to handle and use. Many manufacturers offer high quality digital test instruments with special features, but the basic logic test instruments, LOCO probe, LOCO pulser and LOCO clip given in this book, will be able to find most, if not all, solid failures in the 7400 digital IC series.

Digital test instruments are generally used to detect the presence or absence of a pulse, a train of pulses and/or the static logic condition of a digital circuit. The logic pulser (given in Project 10), is used to inject manually a square wave pulse into a circuit suspected of trouble. The logic probe is used for high or low condition detection and also serves as a pulse stretcher when the pulse is of too short a duration to be seen with the eye. The logic clip (also given in Project 10) will show the logic status of all pins of a DIP IC simultaneously.

The VTVM can be used to test power supply voltages, low output voltages of logic gates resulting from loading effects and current hogging, and to test discrete electronic components.

The VOM can be used the same as the VTVM, but also can measure the current in the digital circuits to determine proper loads and any changes in resistive elements.

The oscilloscope can be used as a DC volt meter, but its main importance is to be able to actually view pulse waveforms and time relationships. With the "scope" you will be able to see if the pulse has deteriorated to a point to cause circuit problems. Some of these pulse defects are preshoot, overshoot, delay in rise time or fall time, too long or too short a pulse width and a delay between input pulses to a logic circuit resulting in faulty operation. A dual-trace "scope" is particularly useful in comparing time relationship of input pulses. Newer digital devices are available that connect to the vertical input of a "scope" which enables you to view several input and output pulses simultaneously.

The pulse generator used for digital work is of the square-edge type and should be adjustable in amplitude, pulse duration (width) and frequency (pulse repetition rate). The amplitudes of the pulses are usually less than 10 volts with a pulse width ranging from 10 nsec to 1 sec. The repetition rates usually go to 100K Hz, but in some cases may go into the megahertz range.

10-2 RECOGNIZING OPEN CIRCUIT FAILURES

An open circuit will affect different circuits, depending upon whether it occurs at an input or output as shown in Figure 10-1. If the open occurs at the input of a logic gate, only that gate will be affected, while the other circuits driven from the same point will be unaffected. Remember, that in TTL a floating input is interpreted as a high level. To detect the open shown in Figure 10-1a, the output of the gate in IC2 would have to be low. Then, a check of the circuit that is driven from this point would reveal that the input to the gate in IC3 would be open because it would be high.

In the case of an open output, all of the circuits driven from that point will be affected as shown in Figure 10-1b. Regardless of whether the output of the gate of IC2 is high or low, all of the driven input circuits to the right of the open would remain permanently high.

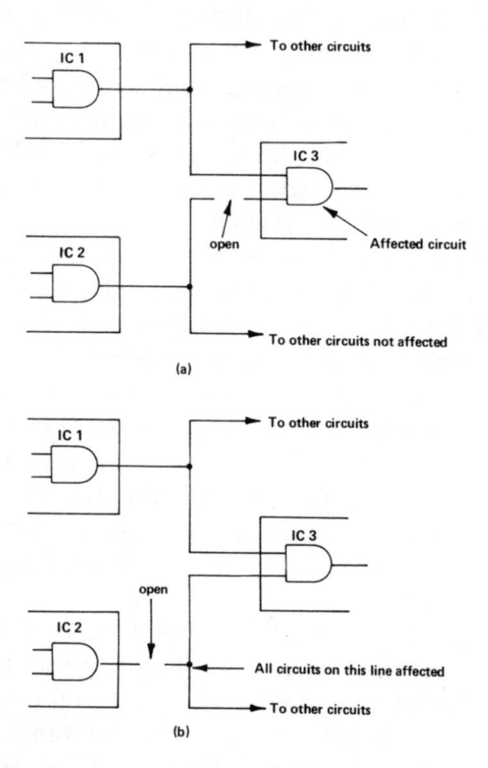

Figure 10-1. Open Circuit Failures (a) Open Input Circuit (b) Open Output Circuit

10-3 RECOGNIZING SHORT-CIRCUIT FAILURES

A short-circuit will affect all of the circuits that are common to the line that is shorted as shown in Figure 10-2. In the case of a short to ground, all of the affected circuits will remain permanently low as shown in Figure 10-2a. If the short occurs to Vcc, all of the affected circuits will remain permanently high as shown in Figure 10-2b. The normal circuit operation beyond these shorted points will usually disappear and this type of failure is usually easy to find.

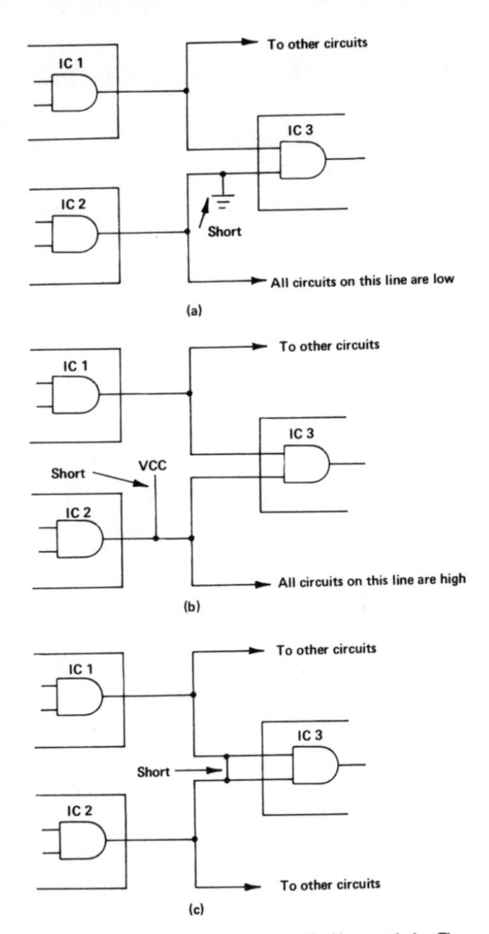

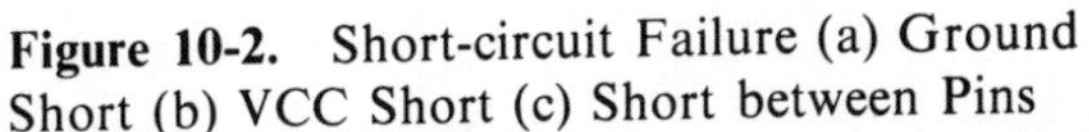
Figure 10-2. Short-circuit Failure (a) Ground Short (b) VCC Short (c) Short between Pins

More difficult to find and analyze is a short between two pins as shown in Figure 10-2c. Whenever the outputs of the gates of IC1 and IC2 go high simultaneously or go low simultaneously, the shorted input pins of IC3 will respond properly. However, if one of the driving outputs goes high, while the other output goes low, the shorted pins will go low. The pull-up transistor in the totem pole output of the driving IC that goes high will supply current to the shorted input pins, but the pull-down transistor in the other totem pole output of the driving IC that goes low will sink this current to ground, thereby, pulling the shorted pins low. The best way to check this type of failure properly is to have one driving output high and the other driving output low and then check the logic condition of the input pins to the driven IC.

10-4 INTERNAL IC FAILURES

Open- and short-circuit failures that occur internally to an IC as shown in Figure 10-3 will have the same effects on circuit operation as the external failures previously mentioned. Internal IC failures sometimes are more difficult to find because it is not possible to make measurements at the inputs and outputs of the logic gates within the IC. For instance, the inputs to the gate of IC3, shown in Figure 10-3a, would appear normal, but the gate would not respond normally because the internal open input would remain at a high level.

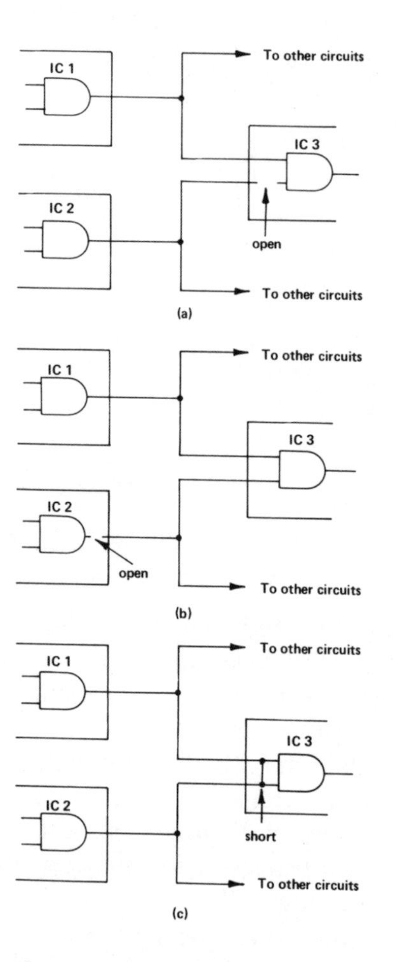

Figure 10-3. Internal IC Open and Short-circuit Failures (a) Open Input (b) Open Output (c) Short between Inputs

In the case of an internal open in the output of a gate as shown in Figure 10-3b, you would have to properly condition the inputs to that gate to make it go low. The output line would then remain high as it would for an external open circuit.

An internal short circuit as shown in Figure 10-3c, whether it be to ground, Vcc or between input pins, would give the same indications as an external short.

Failure of the components within an IC gate as shown in Figure 10-4 will have the effect of keeping the output either permanently high or permanently low. For instance, if Q1 was open, then Q2 would turn on causing Q4 to turn on and the output would be permanently low. If in another case, Q2 was open, Q3 would turn on causing the output to go permanently high.

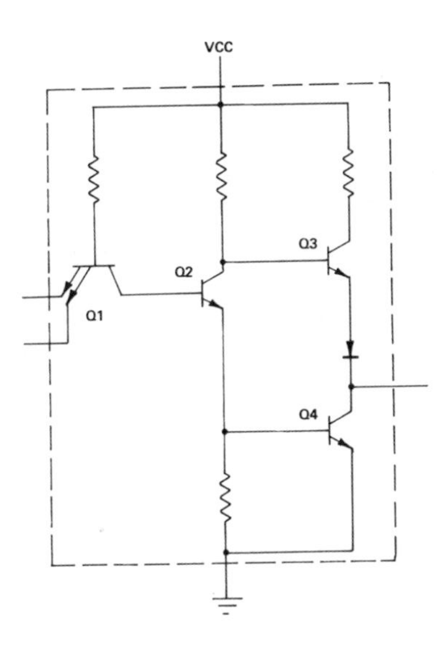

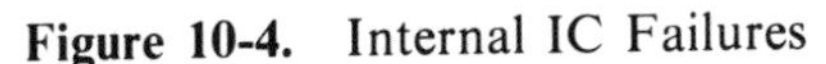

Figure 10-4. Internal IC Failures

10-5 TESTING DIGITAL CIRCUITS

To be able to test digital circuits accurately, a few simple procedures must be followed. When using a manual input pulse device such as the LOCO pulser, the normal inputs to the circuit under test and/or the system clock should be disabled. If the test instruments being used connect to Vcc and ground, such as the LOCO pulser and the LOCO probe, make sure that they are properly connected so that they are protected and ready for use.

To test each input of an individual gate, the other inputs must be connected to the proper level as shown in Figure 10-5a. With this

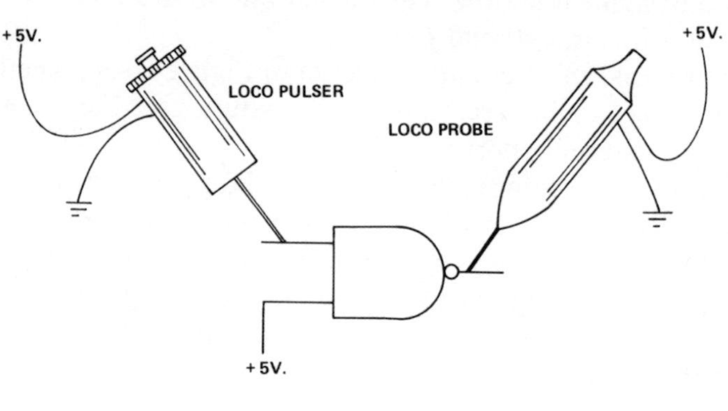

(a)

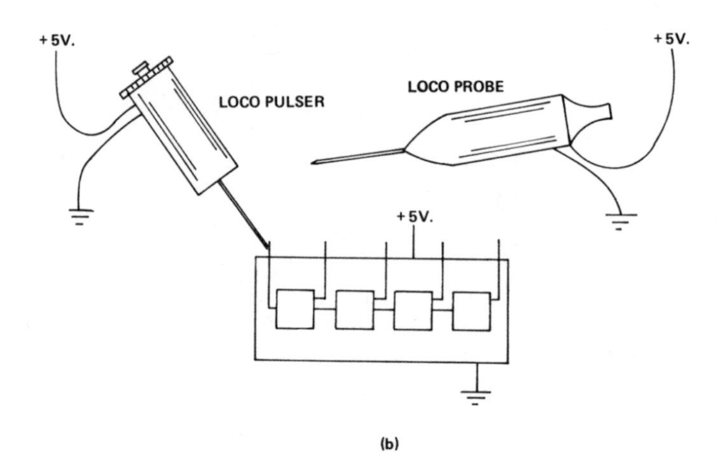

(b)

Figure 10-5. Basic Testing of Digital Circuits
(a) Single-gate Test (b) Counter Testing

NAND gate the input not being tested is tied to a high. The LOCO pulser is placed at the input being tested, while the LOCO probe is placed at the output. The LOCO probe should indicate a high output since the LOCO pulser in the normal condition is low. When the LOCO pulser switch is activated, the input under test should go high and the LED in the LOCO probe will go out indicating a low output. This shows that the input is good, and you can proceed to the next input. Remember to connect the tested input to a high and remove the high from the next input to be tested. Given below is a guide to testing basic logic gates.

Type of Gate	Inputs not under test connected:	Output indication When pulser activated:
AND	high	goes high
OR	low	goes high
Exclusive-OR	low	goes high
NAND	high	goes low
NOR	low	goes low
Inverter	not needed	goes low

The LOCO pulser can be used to enter pulses into a counter under test as shown in Figure 10-5b. Once the desired binary number is entered into the counter, the LOCO probe can be used to check the respective outputs for the corresponding proper high and low conditions.

Testing clocked flip-flops is a little more involved since the various leads are connected to highs and lows. For example, to test a J-K flip-flop for a turn-on operation, the J lead and clear lead would be connected high, while the K lead would be connected low. The LOCO pulser could be used at the clock input and the LOCO probe could then be used to check the Q and $\overline{Q}$ outputs.

More complex ICs are better checked within the digital system as shown in Figure 10-6. Let's say, with this counter-display system that not all of the segments lighted properly for the numbers 0-9. A lamp test input on the decoder/driver IC would help determine if all seven segments were all right. Next, the LOCO pulser could be used

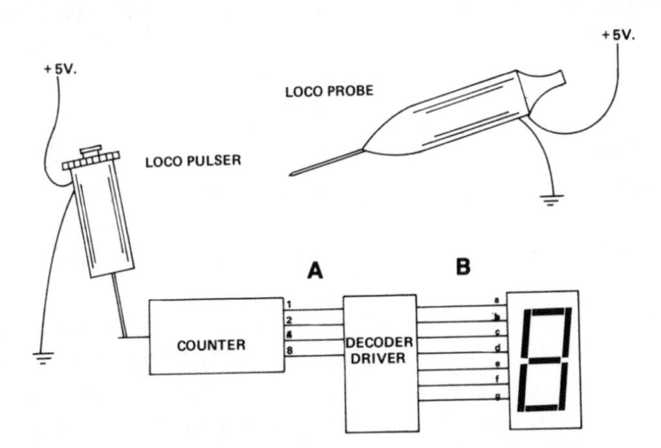

Figure 10-6. Basic Testing of Digital System

at the input of the counter, while the LOCO probe could check the outputs of the counter at point A. This test would determine if the counter was operating properly and that the correct signals were being applied to the decoder/driver inputs. The counter could then be cleared and the count started again, while the decoder/driver outputs were checked with the LOCO probe at point B.

A single PC board could be removed from a highly complex digital system and tested using the LOCO pulser and LOCO probe as shown in Figure 10-7.

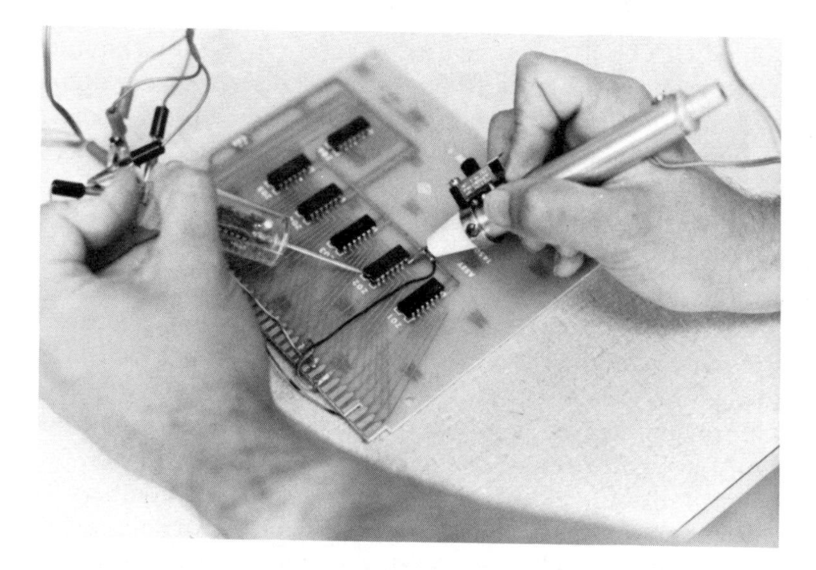

Figure 10-7. Using the LOCO Pulser and the Loco Probe

SUMMARY

Standard electronic test equipment such as the VOM, VTVM, oscilloscope and pulse generator are a must for checking electronic failures. However, digital test equipment, such as the LOCO pulser, LOCO probe and LOCO clip are easier and faster to use when determining digital logic circuit failures.

The essential thing to remember when troubleshooting digital circuits is to make a logic test of the suspected circuit and then apply your knowledge of basic digital circuit operation. The simplicity of testing digital circuits is achieved by entering the proper 1s and 0s and observing the output indications in the form of 1s and 0s.

10-6 PROJECT #10: BUILDING THE "LOCO" PULSER AND THE "LOCO" CLIP

LOCO Pulser

The LOCO pulser uses the same circuit as the bounceless switches constructed in project #4, except that it has a probe for the output and has an LED for a pulse indicator as shown in Figure 10-8a. The output is normally low and will go high to about 3.6 volts when switch S1 is activated. The LED will turn on and is isolated from the output via the double inverters.

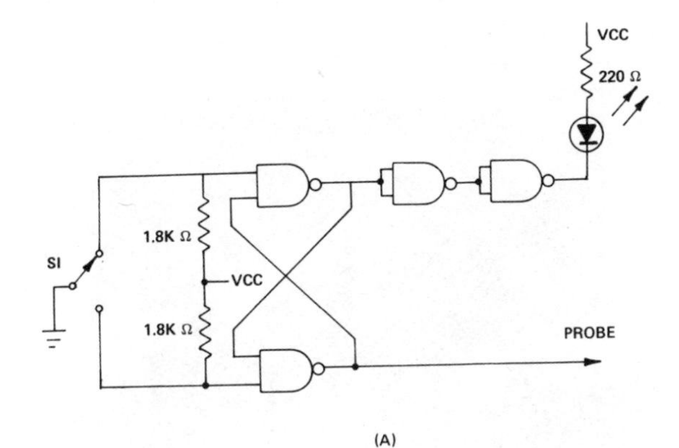

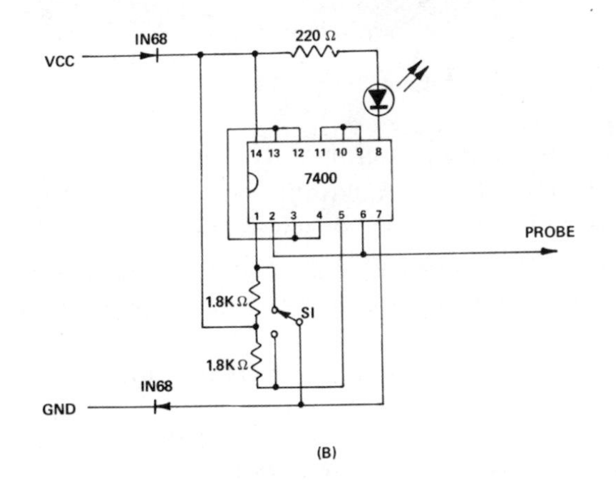

Figure 10-8. LOCO Pulser
(a) Logic Diagram (b) Schematic Diagram

Protection diodes are also used with the LOCO pulser as shown in Figure 10-8b. If the power leads which go to Vcc and ground are connected incorrectly, the diodes are reverse-biased and no current flows to the pulser. When the leads are connected with the proper polarity, the pulser is ready to use. In the low condition, the probe is at about +.7 volts with respect to ground because of the voltage drops of the protection diodes.

The components are mounted on a small perfboard with holes spaced 0.100 inch on centers. After the circuit is wired to the perfboard, it can be placed into a clear plastic medicine bottle as shown in Figure 10-9. The clear plastic permits the LED to be seen without making a special hole in the case. Finally, mini-alligator clips are attached to the power leads.

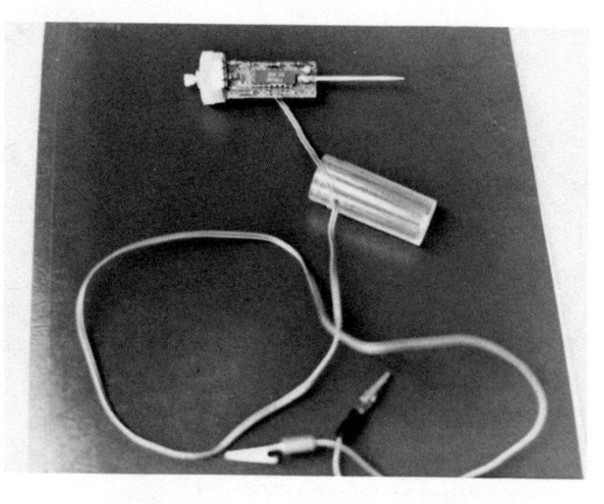

Figure 10-9. LOCO Pulser

To use the LOCO pulser, simply connect the proper power leads to Vcc and ground on the equipment to be tested, then place the probe at the point of the circuit where you want to enter a pulse and press the push-button switch. For a circuit that requires a low trigger pulse, keep the switch pressed and then momentarily release it.

Parts List for LOCO Pulser

1—7400-quad 2-input NAND gate IC
1—220-ohm ¼-watt carbon resistor
2—1.8Kohm ¼-watt carbon resistor
2—1N68 or equivalent diodes
1—Light emitting diode (1.6-2.5 volt)

1—SPDT momentary push-button mini-switch
1—Small medicine bottle or suitable case
2—Mini-alligator clips (color coded red & black)
1—Small perfboard with holes 0.100 inch on center

LOCO Clip

The LOCO clip, like the LOCO probe, is a logic status indicator, but is able to view all pins of a DIP IC simultaneously. A network of diodes is employed that automatically selects Vcc and ground pins, thereby eliminating any special connection requirements. The operation of these diodes is shown in Figure 10-10a. For instance, if pin 1

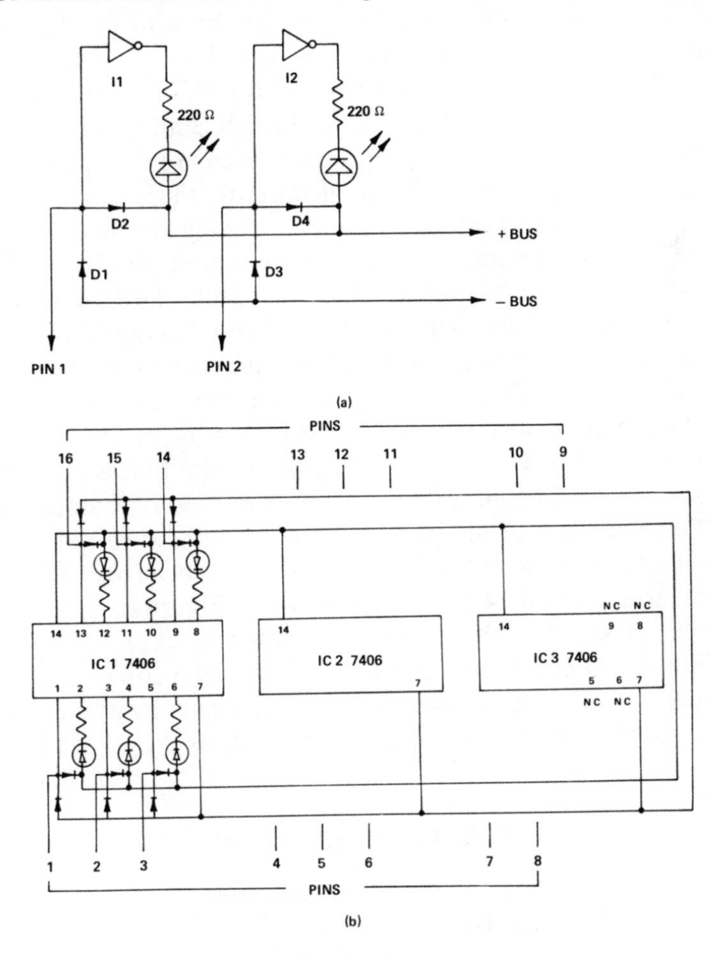

Figure 10-10. LOCO Clip
(a) Diode Operation (b) Schematic Diagram

was connected to Vcc and pin 2 connected to ground, diode D2 would conduct and allow the +bus line to go high. Diode D1 would not conduct and keep the +bus line and −bus line isolated. The −bus line would be established through diode D3, while Diode D4 keeps the +bus line and −bus line isolated from each other in that circuit. Inverter I_1 will see a high at its input and give a low output enabling the LED indicating Vcc to turn on. Inverter I_2 will see a low at its input and gives a high output which keeps the LED indicating ground turned off. The other circuits will respond accordingly to the logic status of the IC being tested.

The individual circuit using one inverter, one 220-ohm resistor, one LED and two diodes is duplicated 15 times to produce a unit capable of viewing 16 pins simultaneously as shown in Figure 10-10b. For simplicity, only IC1 is shown wired, but IC2 and IC3 are wired the same. Only four circuits are needed for IC3 and pins 5,6,9 and 8 are not connected. The 7406 hex inverters used are the open collector type capable of handling 40 ma of current, thereby enabling the LEDs to draw more current without damaging the ICs.

The ICs, diodes and resistors are mounted on the perfboard, that is placed in the bottom of the case, while the LEDs are connected with their anodes to a +bus line mounted in the lid of the case. Wires are then attached from the 220-ohm resistors to the cathodes of the respective LEDs as shown in Figure 10-11a. A 16-lead ribbon cable about 8 inches long connects the inputs of the inverters to the pins on the IC clip. The clip is marked with a dot that is aligned with the index of the IC being tested to give the proper logic status indications. Since the LOCO clip is used to test 14-pin and 16-pin ICs, a second set of numbers are labeled below the upper row of LEDs to make it easier to read 14-pin ICs as shown in Figure 10-11b. The case is a small plastic box that has been painted black.

The LOCO clip is easy to use. Just place the clip over the IC to be tested, making sure the dot on the clip is aligned with the index of the IC, and then read the logic status of the IC as shown in Figure 10-12.

Parts List for LOCO Clip

3—7406 hex inverter buffers/drivers
3—14-pin DIP IC sockets
32—1N60 diodes or equivalent
16—220-ohm ¼-watt carbon resistors
16—Light emitting diodes (1.6-2.5 volt)

1—Small perfboard with holes 0.100 inch on center
1—16-pin IC test clip
1—8-inch long #22 wire 16-lead ribbon cable
1—Small plastic box for case

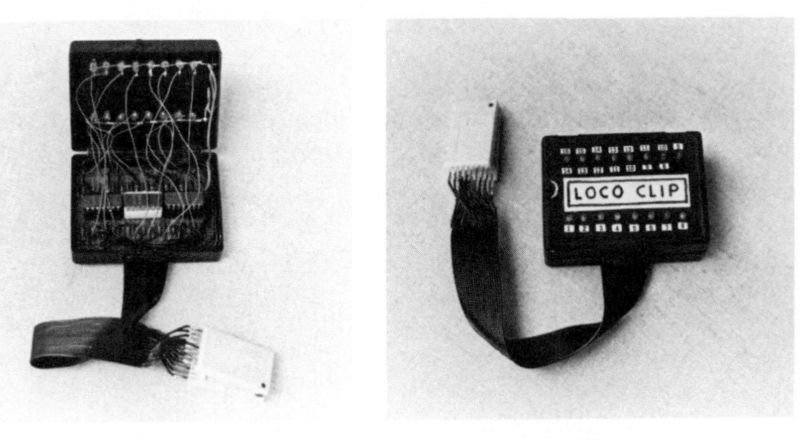

(a) (b)

Figure 10-11. LOCO Clip (a) Open (b) Closed

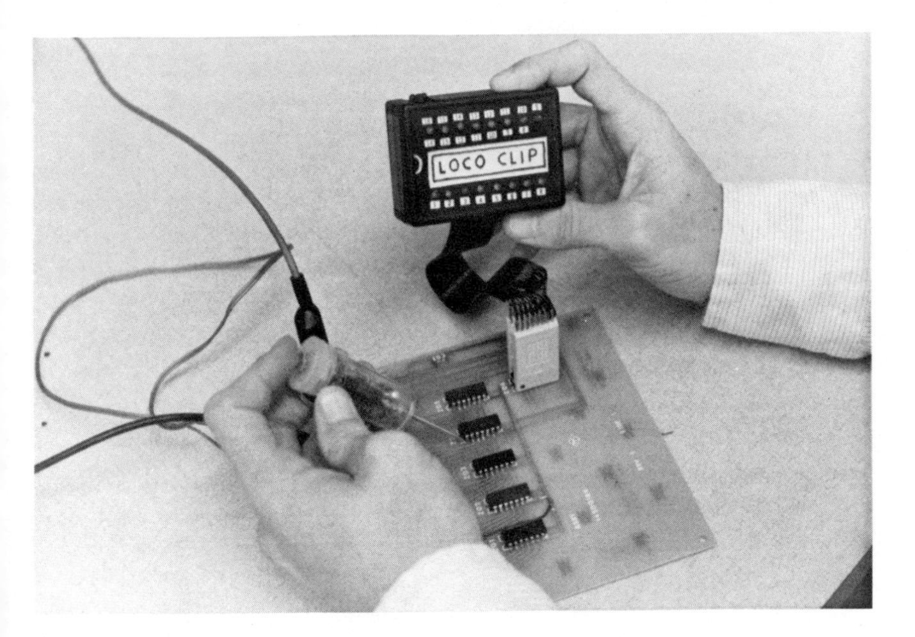

Figure 10-12. LOCO Clip in use

APPENDIX A

Pin configurations for some widely used TTL digital ICs. (See the following list for more information.)

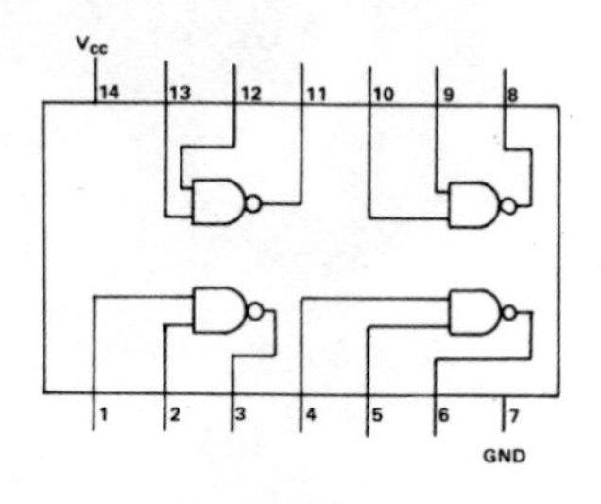

7400

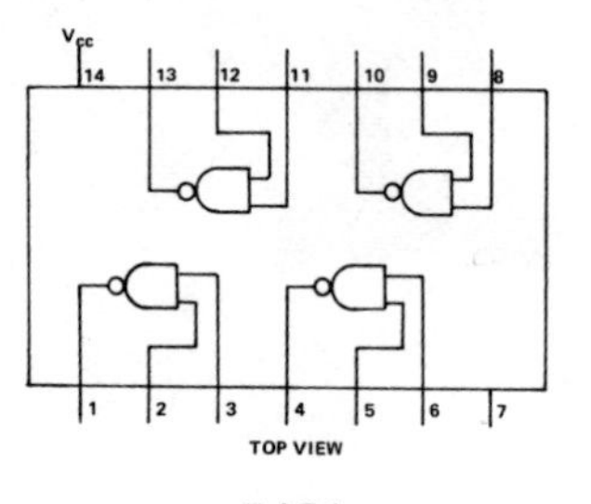

7401

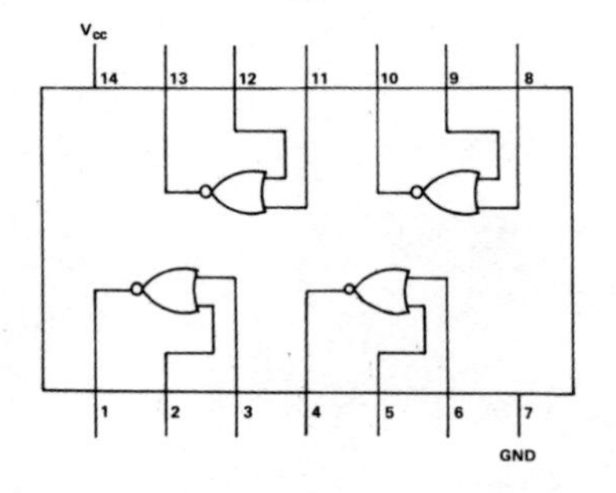

7402

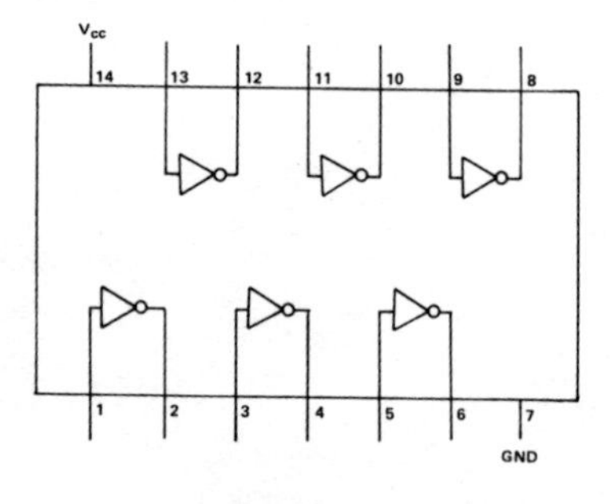

7404

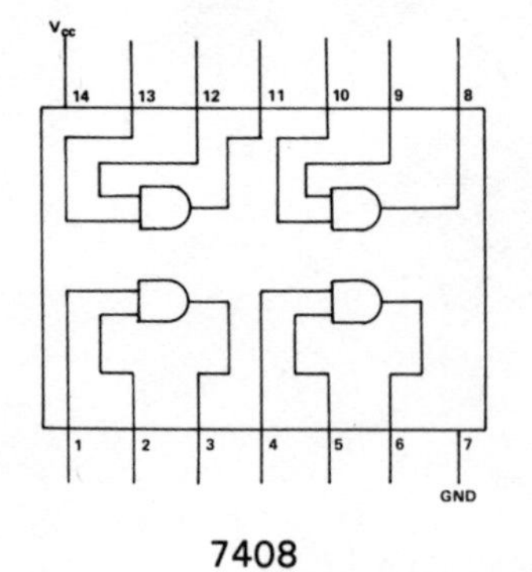

7408

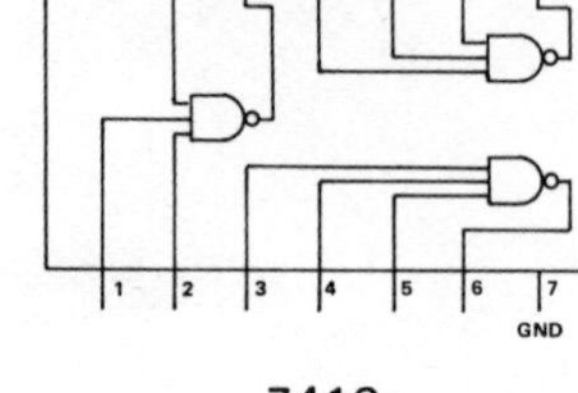

7410

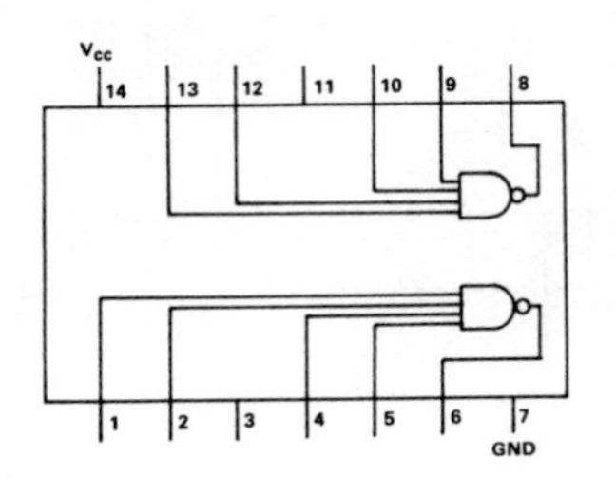

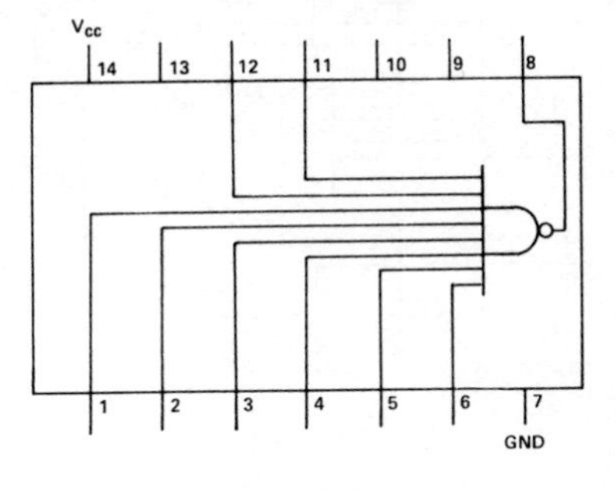

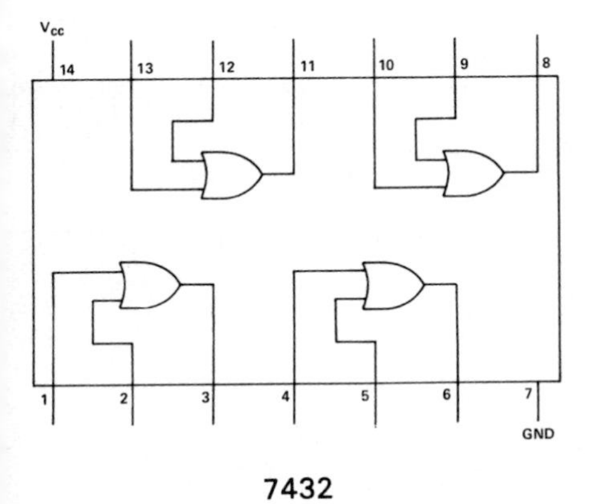

7432

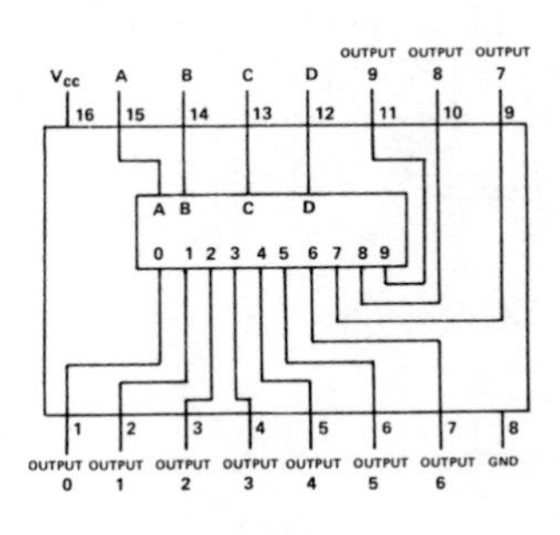

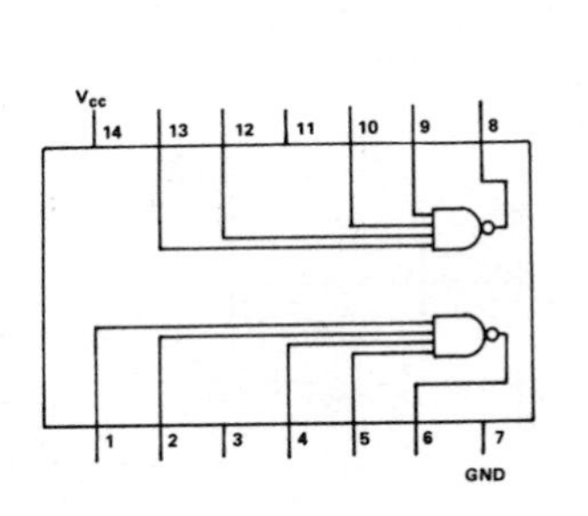

7440

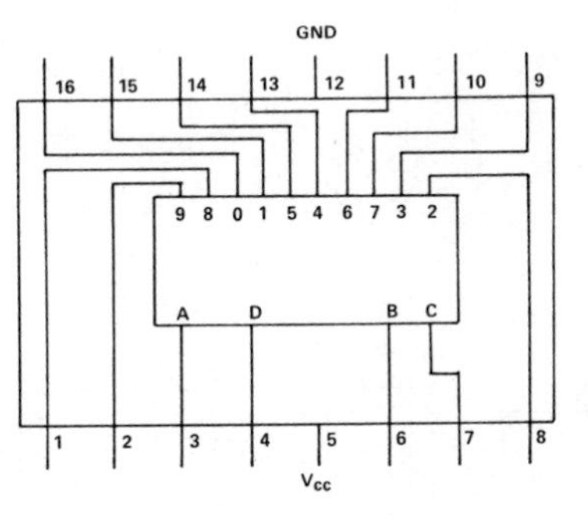

7441

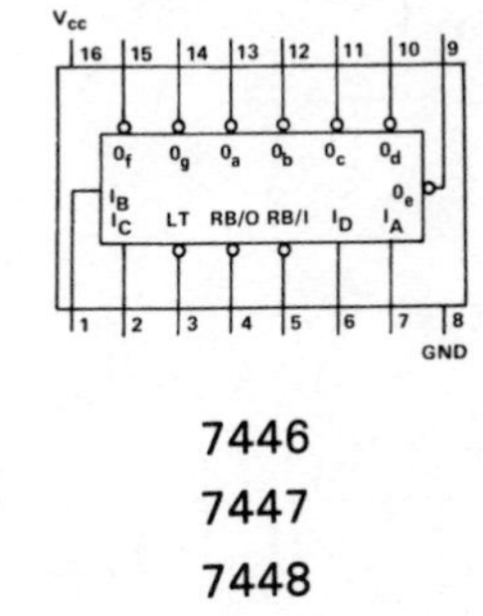

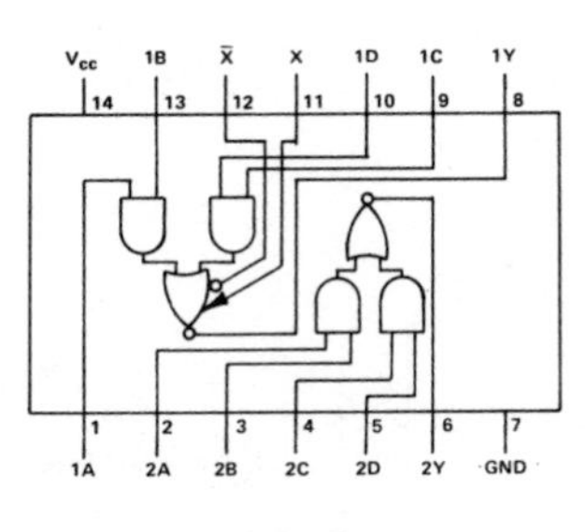

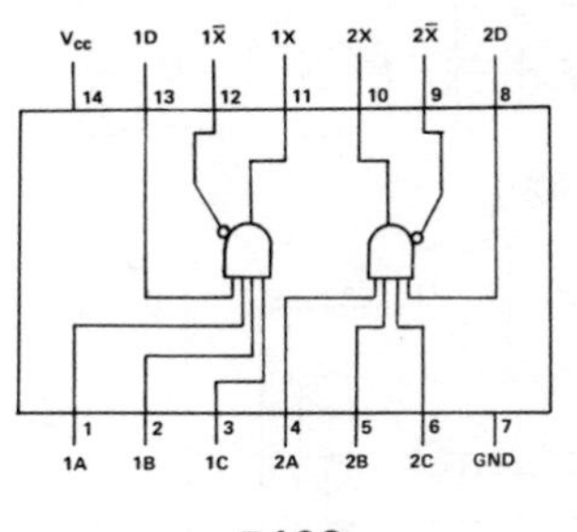

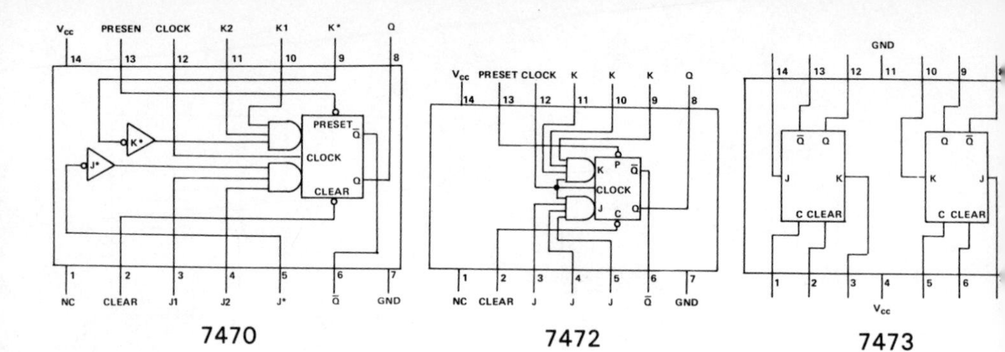

7470

7472

7473

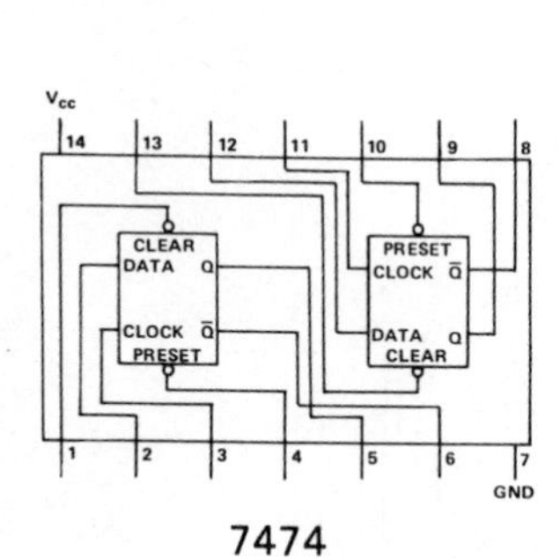

7474

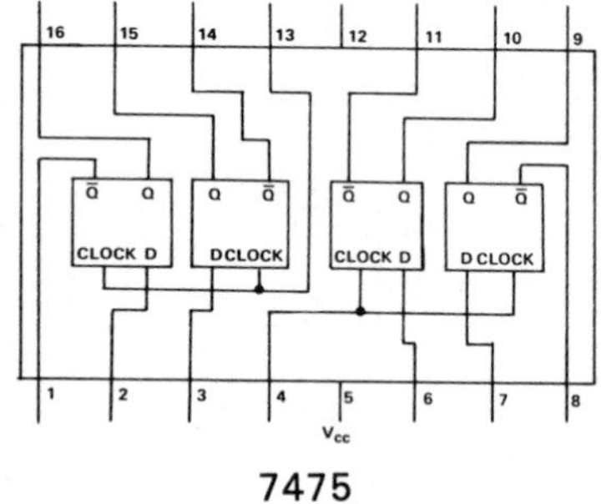

7475

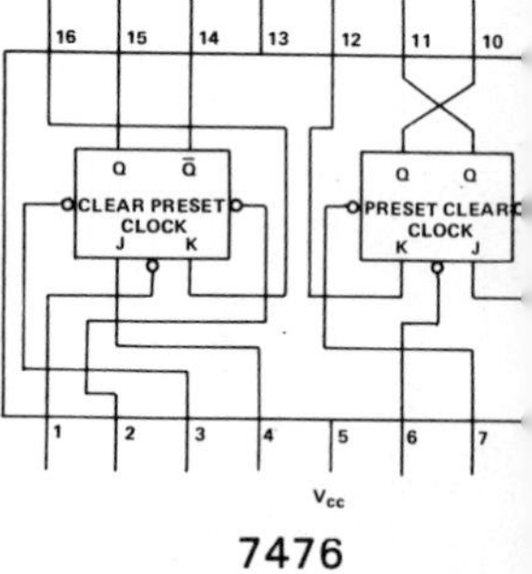

7476

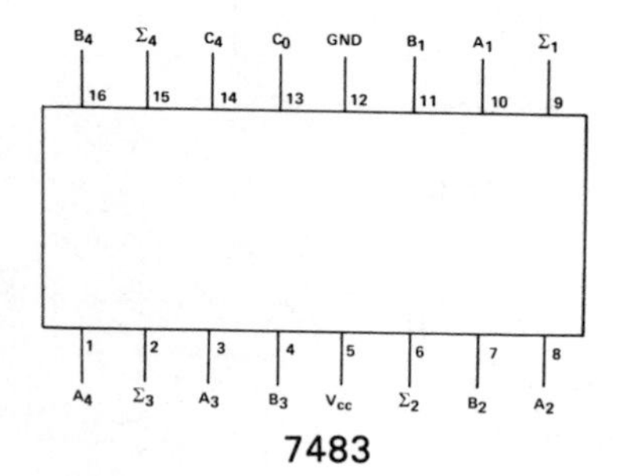

7483

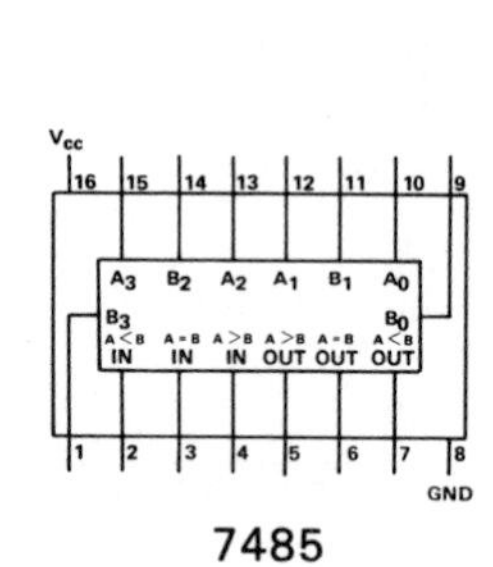

7485

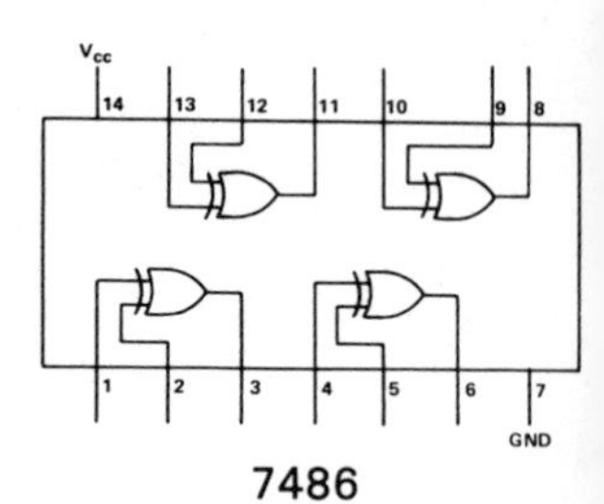

7486

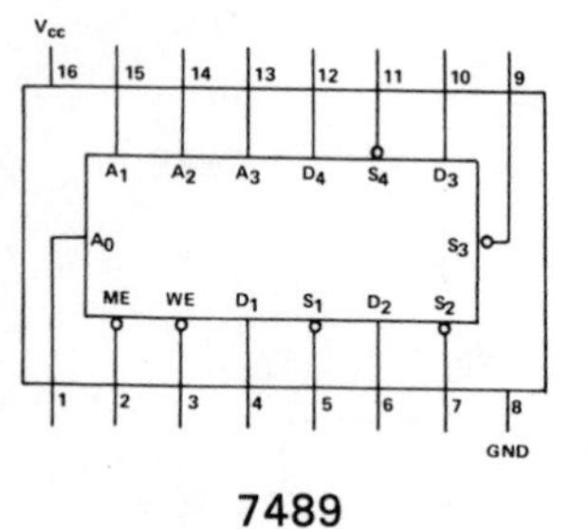

7489

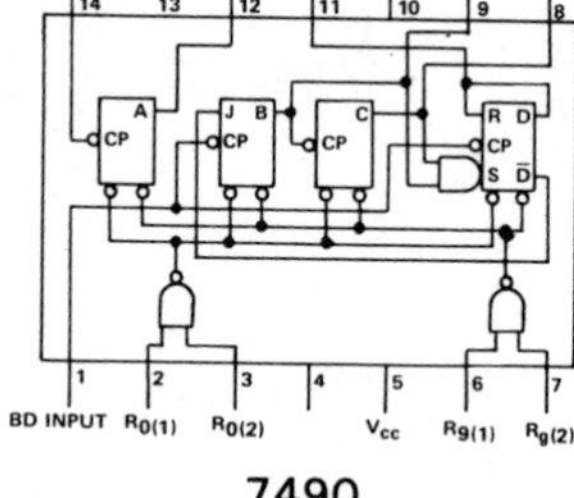

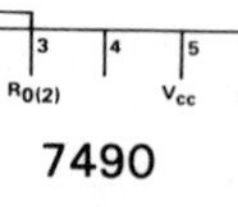

7490

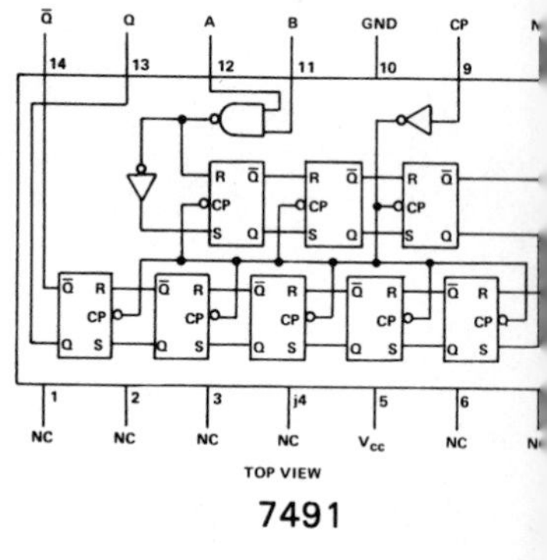

7491

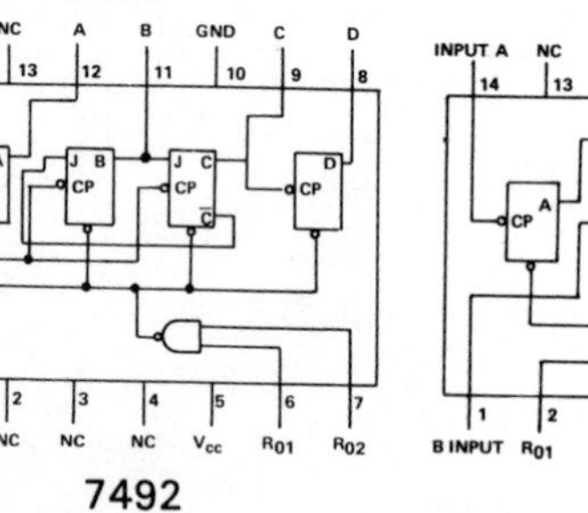

7492

INPUT A NC A D GND B C

B INPUT R01 R02 NC Vcc NC NC

7493

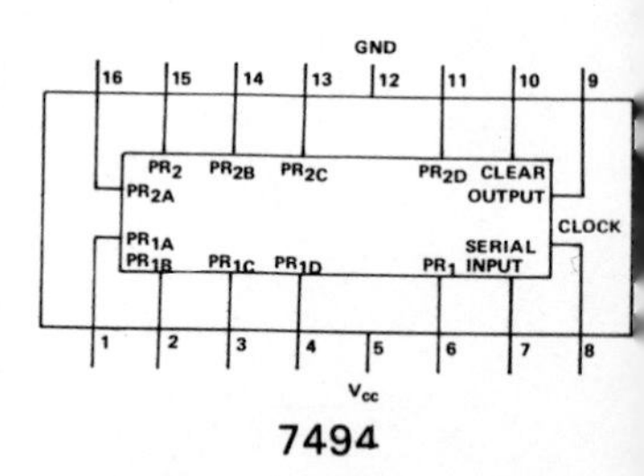

7494

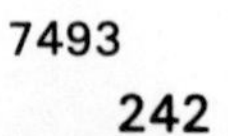

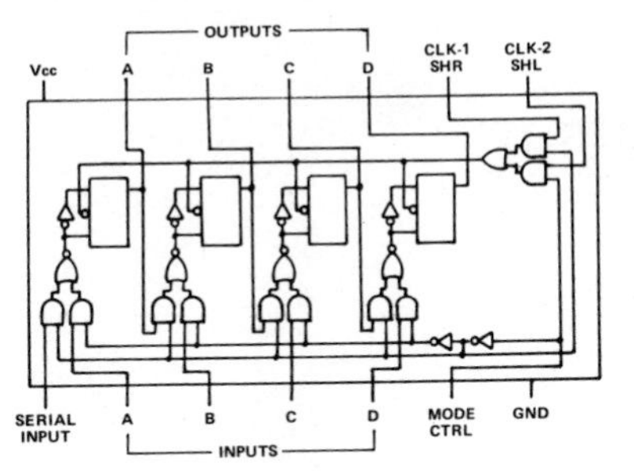

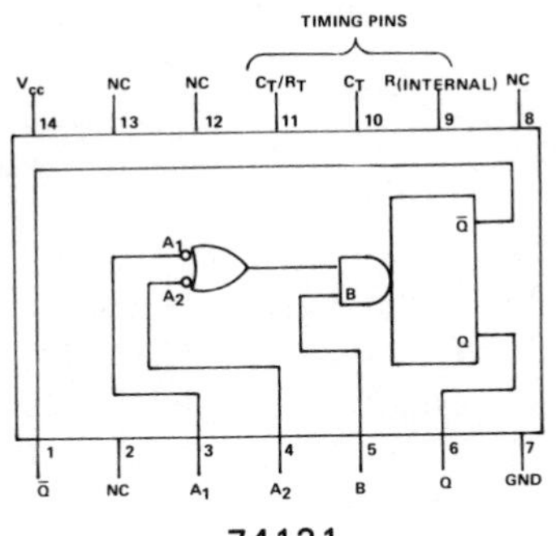

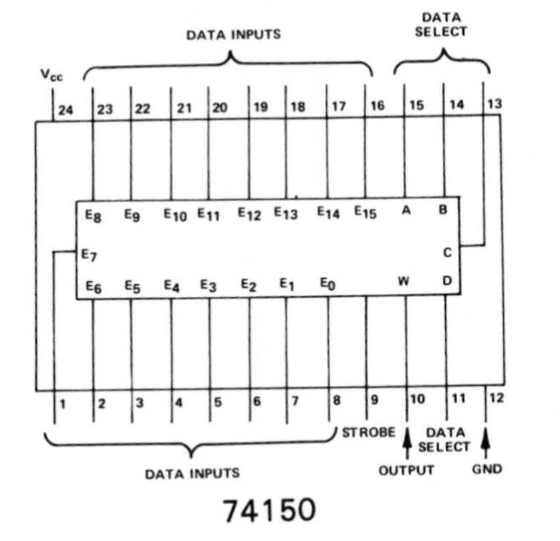

74150

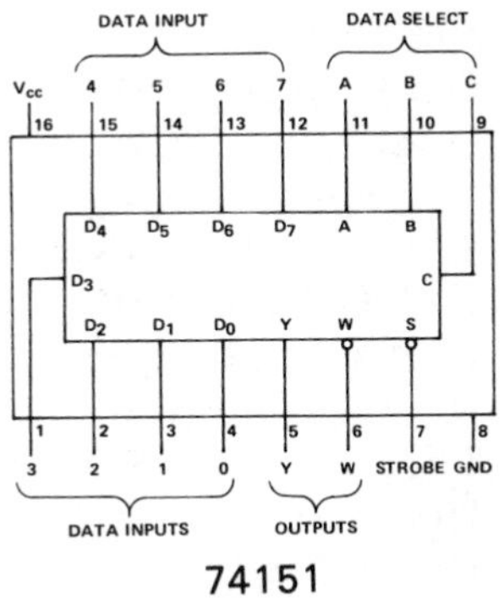

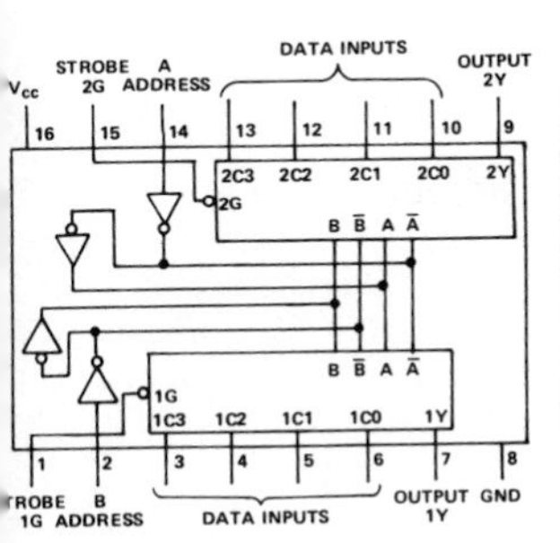

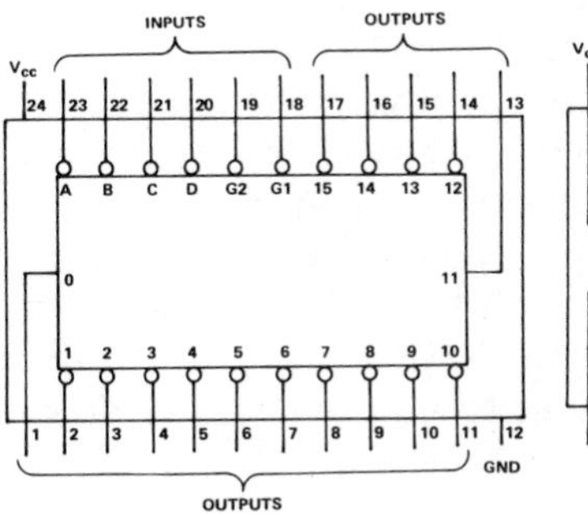

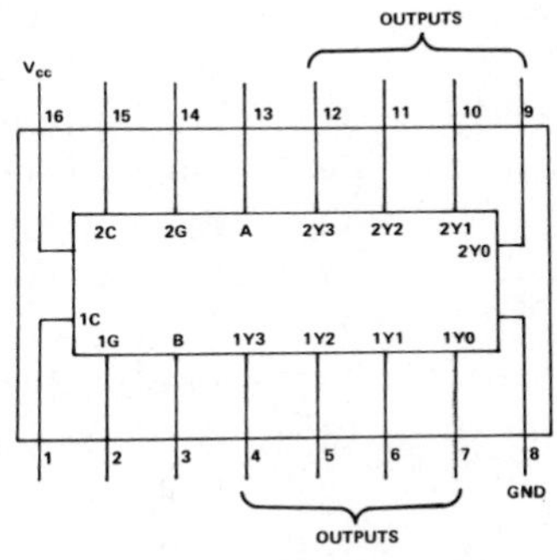

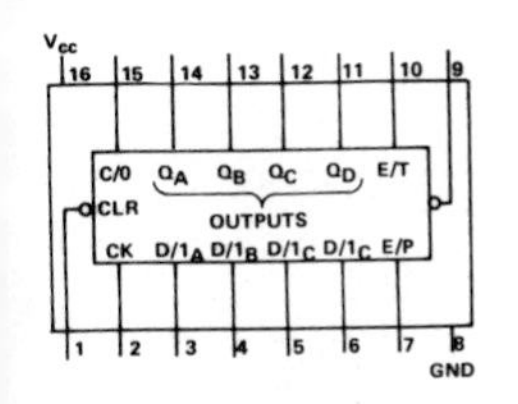

Vcc

16 15 14 13 12 11 10 9

D1 A W/S B W/S W/E R/E Q1

D2 Q2

D3 D4 R/S B R/S A Q4 Q3

1 2 3 4 5 6 7 8

GND

74170

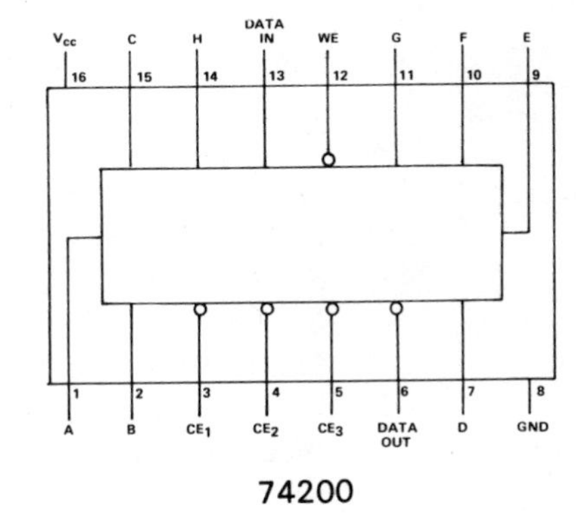

7400 SERIES TTL

7400 Quad 2-Input NAND Gate
7401 Quad 2-Input NAND Gate (O.C.)*
7402 Quad 2-Input NOR Gate
7403 Quad 2-Input NAND Gate (O.C.)
7404 Hex Inverter
7405 Hex Inverter (O.C.)
7406 Hex Inverter Buffer/Drive
7407 Hex Buffer/Driver
7408 Quad 2-Input AND Gate
7409 Quad 2-Input AND Gate (O.C.)
7410 Triple 3-Input NAND Gate
7411 3-Input Positive AND Gate
7412 NAND Gate Inverter
7413 Dual Schmitt-Trigger
7414 Hex Schmitt-Trigger
7415 Triple 3-Input AND Gate (O.C.)
7416 Hex Inverter Buffer/Driver
7417 Hex Buffer/Driver
7418 Triple 3-Input OR Gate
7420 Dual 4-Input NAND Gate
7421 Dual 4-Input Positive AND Gate
7422 Dual 4-Input NAND Gate (O.C.)
7423 Expandable Dual 4-Input NOR Gate with Strobe
7425 Dual 4-Input NOR Gate
7426 Quad 2-Input TTL-MOS Interface Gate
7427 Triple 3-Input NOR Gate
7428 Quad 2-Input NOR Gate with Buffers
7429 Dual 4-Input NOR Gate
7430 8-Input NAND Gate
7432 Quad 2-Input OR Gate
7433 Quad 2-Input NOR Gate with Buffers (O.C.)
7437 Quad 2-Input NAND Buffer
7438 Quad 2-Input NAND Buffer (O.C.)
7439 Quad 2-Input NAND Buffer (O.C.)
7440 Dual 4-Input Buffer
7441 BCD to Decimal Decoder/Nixie TM Driver
7442 BCD to Decimal Decoder
7443 Excess 3-to-Decimal Decoder
7444 Excess 3-Gray-to-Decimal Decoder
7445 BCD to Decimal Decoder/Driver
7446 BCD-to-Seven Seg. Dec./Dr. with 30V Out.

*O.C. stands for open collector

7447	BCD-to-Seven Seg. Dec./Dr. with 15V Out.
7448	BCD-to-Seven Seg. Dec./Driver
7450	Expan. Dual 2-Wide 2-In. AND-OR-INV. Gate
7451	Dual 2-Wide 2-In. AND-OR-Invert Gate
7452	Expandable 7-Wide AND-OR Gate
7453	Expan. 4-Wide 2-In. AND-OR-Invert Gate
7454	4-Wide 2-In. AND-OR-Invert Gate
7455	2-Wide 4-In. AND-OR-Invert Gate
7458	Dual 2-Wide AND-OR-Invert Gate
7459	Dual 2-Wide 2-3 Input AND-OR-Invert Gate
7460	Dual 4-Input Expander
7461	Triple 3-Input Expander
7462	4-Wide AND-OR Expanders
7464	4-2-3-2 Input AND-OR-Invert Gates
7465	4-2-3-2 Input AND-OR-Invert Gates (O.C.)
7470	Edge-Triggered JK Flip Flop
7471	AND-OR Gated J-K Master/Slave FF with Preset
7472	JK Master/Slave Flip Flop
7473	Dual JK Master/Slave Flip Flop
7474	Dual D Flip Flop
7475	Quad D Flip Flop Latch
7476	Dual JK Master/Slave Flip Flop
7477	4-bit Bistable Latch
7478	Dual JK FF with PR, CLR and Common CLK
7480	Gated Full Adder
7481	16-Bit Active-Element Memory
7482	2-Bit Binary Full Adder
7483	4-Bit Binary Full Adder
7484	16-Bit Active-Element Memory
7485	4-Bit Magnitude Comparator
7486	Quad EXCLUSIVE-OR Gate
7487	4-Bit True/Complement-Zero/One Element
7488	256-Bit Read Only Memory
7489	64-Bit Read Only Memory
7490	Decade Counter
7491	8-Bit Random Access Read/Write Memory Shift Register
7492	Divide by 12 Counter
7493	4-Bit Binary Counter
7494	4-Bit Shift Register (Parallel-IN, Serial Out)
7495	4-Bit Right Shift/Left Shift Register
7496	5-Bit Parallel-In/Parallel-Out Shift Reg.
7497	Synchronous 6-Bit Binary Rate Multiplier
7498	4-Bit Data Selector/Storage Register
7499	4-Bit Right-Shift Left-Shift Register
74100	4-Bit Bistable Latch
74101	AND-OR Gated JK Neg. Edge Triggered FF with PR
74102	AND Gated JK Neg. Edge Triggered FF with PR & CLR

74103 Dual JK Neg. Edge Triggered FF with CLR
74104 (29000) Gated JK Master/Slave FF
74105 (29001) Gated JK ($\overline{JK}$) Master/Slave FF
74106 Dual JK Neg. Edge Triggered FF with PR & CLR
74107 Dual JK Master/Slave Flip Flop
74108 Dual JK Neg. Edge Triggered FF with PR, CLR, & CLK
74109 Dual JK Pos. Edge Triggered FF with PR & CLR
74110 AND-Gated JK Master/Slave FF with Data Lockout
74111 Dual JK Master/Slave FF with Data Lockout
74112 Dual JK Neg. Edge Triggered FF with PR & CLR
74113 Dual JK Neg. Edge Triggered FF with PR
74114 Dual JK Neg. Edge Triggered FF with PR, CLR & CLK
74116 Dual 4-Bit Latches with CLR
74120 Dual Pulse Synchornizers/Drivers
74121 Monostable Multivibrator
74122 Retriggerable Monostable Multi-vibrator W. Clear
74123 TTL/Monostable Multivibrator
74125 TRI-STATE Quad Buffer
74126 TRI-STATE Quad Buffer
74128 50 ohm Line Driver
74132 Quad Schmitt Trigger
74133 13-Input POS. NAND Gate
74134 12-Input POS NAND Gate with 3-State Output
74135 Quad EXclusive OR/NOR Gates
74136 Quad EXclusive OR/NOR Gates (O.C.)
74138 3-to-8 Line Decoders/Demultiplexers
74139 2-to-4 Line Decoders/Demultiplexers
74140 Dual 4-Input NAND/50 ohm Line Drivers
74141 BCD to Decimal Decoder/Driver
74142 Center Latch Nixie
74143 Center Decoder
74144 Center Decoder
74145 BCD to Decimal Decoder/Driver
74147 Decimal-to-Binary Encoder
74148 Octal-to-Binary Encoder
74150 16-Line to 1-Line Multiplexer
74151 8-Channel Digital Multiplexer
74152 8-to-1 Line Data Selector/Multiplexer
74153 Dual 4:1 Multiplexer
74154 4-Line to 16-Line Decoder/Demultiplexer
74155 Dual 2:4 Demultiplexer
74156 Dual 2:4 Demultiplexer
74157 Quad 2-Input Multiplexer

74158 Quad 2-to-1 Line Data Selector/Multiplexer
74159 4-to-16 Line Decoders/Demultiplexers (O.C.)
74160 Decade Counter with Asynchronous Clear
74161 Synchronous 4-Bit Counter
74162 Decade Counter with Asynchronous CLR
74163 Synchronous 4-Bit Counter
74164 8-Bit Serial-in Parallel-out Shift Register
74165 8-Bit Parallel-in Serial-out Shift Register
74166 8-Bit Shift Register
74167 Rate Multiplier
74170 4 × 4 Register File
74172 Register File
74173 TRI-STATE Quad D FF
74174 Hex D Flip Flop W/Clear
74175 Quad D Flip Flop W/Clear
74176 Presettable Decade Counter
74177 Presettable Binary Counter
74178 4-Bit Parallel-access Shift Register
74179 4-Bit Parallel-access Shift Register with CLR & $\overline{Q}$ Out
74180 8-Bit Odd/Even Parity Gener./Chec.
74181 Arithmetic Logic Unit
74182 Look-Ahead Carry Generator
74183 Dual Carry-Save Full Adders
74184 BCD to Binary Converter
74185 Binary to BCD Converter
74186 512-Bit PROM
74187 1024-Bit Read Only Memory
74188 256-Bit PROM
74190 Up-Down Decade Counter
74192 Up-Down Decade Counter
74193 Up-Down Binary Counter
74194 4-Bit Bit Bidirectional Universal Shift Register
74195 4-Bit Parallel-Access Shift Reg.
74196 Presettable Decade Counter
74197 Presettable Binary Counter
74198 8-Bit Shift Register
74199 8-Bit Shift Register
74200 Tri-State 256-Bit Random Access Memory
74251 Tri-State 74151
74260 Dual 5-Input Pos. NOR Gates
74279 Quad $\overline{S}$ $\overline{R}$ Latches
74284 Tri-State Four Bit Multiplexer
74285 Tri-State Four Bit Multiplexer

APPENDIX B

The following is a partial list of leading suppliers of integrated circuits and digital electronic devices.

Active Electronic Sales Corp.
P. O. Box 1035
Framingham, MA 01701

Altaj Electronics
P. O. Box 38544
Dallas, TX 75238

Ancrona Corp.
P. O. Box 2208R
Culver City, CA 90230

Babylon Electronics
P. O. Box 41778
Sacramento, CA 95841

DiGi-Key Corporation
P. O. Box 126
Thief River Falls, MN 56701

International Electronics Unlimited
P. O. Box 1708
Monterey, CA 93940

James Electronics
P. O. Box 822
Belmont, CA 94002

New Electronics
P. O. Box 7248
Carmel, CA 93921

OEMorsco
2403 Charleston Road
Mountain View, CA 94043

Olson Electronics
260 S. Forge Street
Akron, OH 44327

Poly Paks
P. O. Box 942
South Lynnfield, MA 01940

Radio Shack
2615 West 7th Street
Fort Worth, TX 76107

Solid State Sales
P. O. Box 74A
Somerville, MA 02143

Solid State Systems, Inc.
P. O. Box 617
Columbia, MO 65201

APPENDIX C

The following is a partial list of leading manufacturers of integrated circuits and digital electronics devices.

Advanced Micro Devices
901 Thompson Place
Sunnyvale, CA 94086

Burroughs Corporation
Electronic Components Division
P. O. Box1226
Plainfield, N. J. 07061

Fairchild Semiconductor
313Fairchild Drive
Mountain View, CA 94040

Harris Semiconductor
Box 883
Melbourne, FL 32901

Hewlett-Packard Co.
1501 Page Mill Road
Palo Alto, CA 94306

Intel Corporation
3065 Bowers Avenue
Santa Clara, CA 95051

Intersil, Inc.
10900 N. Tantau Avenue
Cupertino, CA 95014

ITT Semiconductors
3301 Electronics Way
West Palm Beach, FL 33407

Litronix
19000 Homestead Road
Cupertino, CA 95014

Monsanto
Electronics Division
3400 Hillview Avenue
Palo Alto, CA 94304

Motorola Semiconductor Products
Box 20912
Phoenix, AZ 85036

National Semiconductor Corp.
2900 Semiconductor Drive
Santa Clara, CA 95051

Raytheon Semiconductor
350 Ellis Street
Mountain View, CA 94040

RCA Solid State Division
Route 202
Somerville, NJ 08876

Signetics
811 E. Argues Ave.
Sunnyvale, CA 94086

Texas Instruments
Box 5012
Dallas, TX 75222

INDEX

N

O

P

R